Thomas Mestekemper

Energy Demand Forecasts and Dynamic Water Temperature Management

Thomas Mestekemper

Energy Demand Forecasts and Dynamic Water Temperature Management

Improving Multivariate Forecasting through Modern Statistical Modelling

Südwestdeutscher Verlag für Hochschulschriften

Impressum/Imprint (nur für Deutschland/only for Germany)
Bibliografische Information der Deutschen Nationalbibliothek: Die Deutsche Nationalbibliothek verzeichnet diese Publikation in der Deutschen Nationalbibliografie; detaillierte bibliografische Daten sind im Internet über http://dnb.d-nb.de abrufbar.
Alle in diesem Buch genannten Marken und Produktnamen unterliegen warenzeichen-, marken- oder patentrechtlichem Schutz bzw. sind Warenzeichen oder eingetragene Warenzeichen der jeweiligen Inhaber. Die Wiedergabe von Marken, Produktnamen, Gebrauchsnamen, Handelsnamen, Warenbezeichnungen u.s.w. in diesem Werk berechtigt auch ohne besondere Kennzeichnung nicht zu der Annahme, dass solche Namen im Sinne der Warenzeichen- und Markenschutzgesetzgebung als frei zu betrachten wären und daher von jedermann benutzt werden dürften.

Verlag: Südwestdeutscher Verlag für Hochschulschriften GmbH & Co. KG
Dudweiler Landstr. 99, 66123 Saarbrücken, Deutschland
Telefon +49 681 37 20 271-1, Telefax +49 681 37 20 271-0
Email: info@svh-verlag.de

Zugl.: Bielefeld, Bielefeld University, Diss., 2011

Herstellung in Deutschland:
Schaltungsdienst Lange o.H.G., Berlin
Books on Demand GmbH, Norderstedt
Reha GmbH, Saarbrücken
Amazon Distribution GmbH, Leipzig
ISBN: 978-3-8381-2782-8

Imprint (only for USA, GB)
Bibliographic information published by the Deutsche Nationalbibliothek: The Deutsche Nationalbibliothek lists this publication in the Deutsche Nationalbibliografie; detailed bibliographic data are available in the Internet at http://dnb.d-nb.de.
Any brand names and product names mentioned in this book are subject to trademark, brand or patent protection and are trademarks or registered trademarks of their respective holders. The use of brand names, product names, common names, trade names, product descriptions etc. even without a particular marking in this works is in no way to be construed to mean that such names may be regarded as unrestricted in respect of trademark and brand protection legislation and could thus be used by anyone.

Publisher: Südwestdeutscher Verlag für Hochschulschriften GmbH & Co. KG
Dudweiler Landstr. 99, 66123 Saarbrücken, Germany
Phone +49 681 37 20 271-1, Fax +49 681 37 20 271-0
Email: info@svh-verlag.de

Printed in the U.S.A.
Printed in the U.K. by (see last page)
ISBN: 978-3-8381-2782-8

Copyright © 2011 by the author and Südwestdeutscher Verlag für Hochschulschriften GmbH & Co. KG and licensors
All rights reserved. Saarbrücken 2011

To Ida Kathleen

Contents

1	**Introduction**	**1**
	1.1 Dynamic Temperature Management in the River Wupper	2
	1.2 Outline of this Thesis	6
2	**Theoretical Background**	**8**
	2.1 Generalized Additive Models	8
	2.1.1 Penalized Splines	10
	2.1.2 Representing GAMs as Generalized Linear Mixed Model	13
	2.1.3 Estimation	15
	2.1.4 Spline Bases and Penalties	17
	2.1.5 Computational Issues	19
	2.2 The EM-Algorithm	20
	2.3 Curve Registration	21
	2.3.1 History and Recent Advances	21
	2.3.2 Landmark-based Curve Registration	24
	2.3.3 Monotone Smoothing with Quadratic Programming	27
	2.4 Approximate Dynamic Factor Models	29
	2.4.1 History and Recent Advances	29
	2.4.2 Common Factor Estimation via Principal Components Analysis .	33
	2.4.3 Common Factor Estimation via Generalized Principal Components	35
	2.4.4 Principal Components Analysis vs. Exploratory Factor Analysis .	36

Contents

3 Application: Landmark Specification in Water Temperature Data 39
- 3.1 Landmark Specification 42
  - 3.1.1 Running Means 43
  - 3.1.2 Temperature Thresholds 43
  - 3.1.3 Daily Temperature Curve 47
  - 3.1.4 Correlation between Water and Air Temperature 48
- 3.2 Registering the Data 52
  - 3.2.1 Landmark-based Time-Warping 52
  - 3.2.2 Linking to Ecological Data 55
- 3.3 Variability of Landmarks 57
  - 3.3.1 The Bootstrap Procedure 57
  - 3.3.2 Bootstrap Application 58
- 3.4 Results 61

4 Application: Forecasting Water Temperature with Dynamic Factor Models 62
- 4.1 Models and Estimation 65
  - 4.1.1 Removing the Seasonal Component from the Data 65
  - 4.1.2 The Dynamic Factor Model 67
  - 4.1.3 The Benchmark Model 75
- 4.2 Forecasting 76
  - 4.2.1 One Day Ahead Forecast 76
  - 4.2.2 Longer Forecasting Horizons 77
  - 4.2.3 Forecasting Performance 78
  - 4.2.4 Forecasting Errors 79
- 4.3 Application to the Dataset 80
  - 4.3.1 Forecasting Water Temperature with Dynamic Factor Models . . 81
  - 4.3.2 Comparison to the Benchmark Model 87
  - 4.3.3 Comparison to Other Modellling Approaches in Hydrology 88
- 4.4 Data Quality 90
- 4.5 Results 93

Contents

5 Application: Forecasting Energy Demand with Dynamic Factor Models   94
    5.1 Time Series Component 98
          5.1.1 Periodic Autoregression 98
          5.1.2 Dynamic Factor Models 100
    5.2 Mean Component 104
    5.3 Wuppertal District Heat Demand 108
          5.3.1 Mean Component 108
          5.3.2 Time Series Components & Forecasts 109
    5.4 Victorian Electricity Demand 115
          5.4.1 Mean Component 115
          5.4.2 Time Series Component & Forecasts 119
    5.5 Results 123

6 Discussion and Perspective   124

Literature   126

1 Introduction

Due to the technological advances in the last thirty years more and more data become available. Scientists from many fields are monitoring huge numbers of variables that might be univariate, multivariate, longitudinal or even functional. This leads to an exploding numerical complexity when conclusions shall be drawn from this data or interdependencies shall be analyzed within the framework of a regression setting. Often classical statistical tools fail to provide reliable results. A growing number of unobservable parameters has to be estimated which increases the amount of parameter uncertainty that is contained in the model. Therefore data compression methods have received much attention in statistics during the last three decades. Exploratory factor analysis and principal components analysis that were formerly well known in sociology and psychology have gained importance in other fields like econometry, for example. Ramsay & Silverman (2005) coined the term "functional data analysis" and developed multivariate techniques for functional data, i.e. for discretisized versions of functional observations that are a posteriori smoothed with the help of a spline basis and an appropriate smoothness penalty. In the economic field where time series forecasts are of great interest Geweke (1977) and Sargent & Sims (1977) suggested the use of so-called "dynamic factor models" that reduce the dimension of a high-dimensional forecasting problem thereby making it tractable for classical time series techniques. In the recent past dynamic factor models have gained popularity and found applications in many fields. In this thesis we look at high-dimensional data sets and present new variations and applications of the methodologies mentioned above.

1 Introduction

1.1 Dynamic Temperature Management in the River Wupper

The applications presented in this thesis are related to a research project authorized by the municipal utility of the city of Wuppertal in cooperation with the *Wupperverband*. These institutions supplied a dataset which included the following covariates for different time spans on an hourly basis: water and air temperature, stream flow, precipitation, global radiation, heat demand. The measurements were partially taken at different locations in the area of interest. The river Wupper is located in the north-western part of Germany and embouches near Cologne into the Rhine. The Wuppertal municipal utility operates two fuel based power plants on its banks which use river water as cooling device and by doing so heat up the stream. This has crucial impacts on the ecological conditions, for example on the amount of dissolved oxygen in the water which in turn determines the species that can inhabit the river. Naturally, the Wupper belongs to the waterbodies which are inhabited by Salmonidae and further downstream by Cyprinidae. If the water temperature in certain states of development exceeds a given threshold the spawning cycle of both families of fish can be severely disturbed with the long term effect that the prefered species become extinct. To avoid effects like that the Water Framework Directive of the European Union (see directives 2000/60/EC and 2005/646/EC) obliges the member states to achieve good qualitative and quantitative status of waterbodies until 2015.

If the water temperature reaches critical values there are mainly two solutions to avoid crossing the corresponding threshold. Firstly, the stream water can be cooled down by mixing in colder water taken from an upstream water reservoir but in long heat periods this would waste the reserves for the year within a short period of time. Secondly, the warming caused by the power plants can be lowered by throttling or even shutting down the gas engines. This in turn causes costs for not being able to run the engines at the desired level. For economical reasons two types of forecasts are required which shall be provided by applying various statistical tools in the remainder of this paper:

1 Introduction

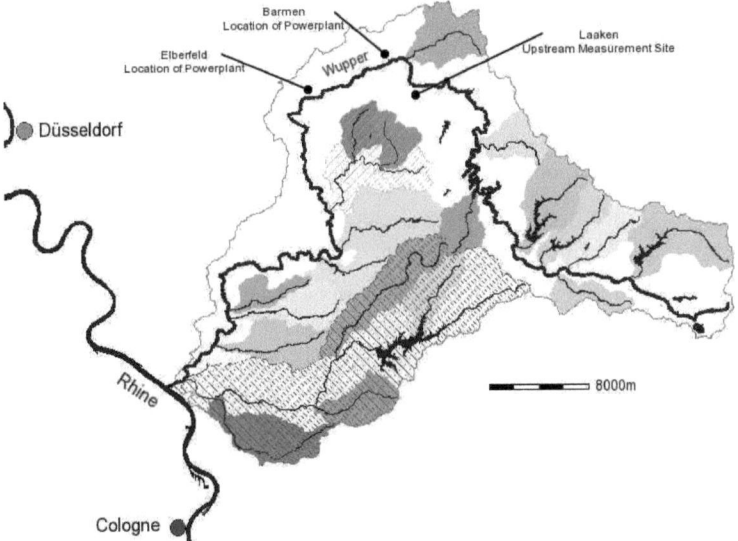

Figure 1.1: Map of the drainage area of the river Wupper. Points of interest are marked by black spots. All three locations belong to the city of Wuppertal (Source: www.wupperverband.de).

1 Introduction

1. A **water temperature forecast** at an upstream location close to the first of the two power plants to determine which amount of heat can be dissipated. The locations of the measurement site "Laaken" and the power plants are indicated in Figure 1.1 which shows the entire drainage area of the river Wupper. A methodology for forecasting water temperature using dynamic factor models is presented in Chapter 4.

2. A **heat demand forecast** which leads to an expected amount of waste heat that has to be dissipated using stream water. Of course there is also an economical interest in such kind of a forecast. A procedure for forecasting heat demand using dynamic factor models is described in Chapter 5.

Knowledge about future values of both, water temperature and waste heat amount, allows to assess if in the following days the corresponding water temperature threshold is at risk in which case the engine output has to be reduced. Furthermore, it is desirable to provide both forecasts on a fine resolution as energy is traded in short time intervals at the European Energy Exchange in Leipzig, Germany (www.eex.com).

A third aspect related to economic worthwhile water temperature management is to fix the appropriate threshold for the current season. Figure 1.2 illustrates the relevant thresholds for the Brown Trout. In the beginning of winter Salmonids start spawning and a water temperature up to 10°C is acceptable. In spring when hatching is finished the threshold can be risen to 12°C at first and then 14°C. These thresholds are allowed to be crossed once in a while but warming up the water over a longer period causes death of the young fish. In summer there is a strict maximum temperature of 25°C which must not be exceeded because this would cause death even to adult fish as its proteins begin to clot. In autumn the river water temperature should be reduced stepwisely from 14°C to 10°C to trigger the spawning cycle, again.

The above mentioned temperatures are given for an average year but in practise it is very implausible to assume that the thresholds are attached to certain timepoints. The beginning of the winter which is the most important pivot in the spawning cycle of fish can come earlier in some years and later in others. The question arises how to

1 Introduction

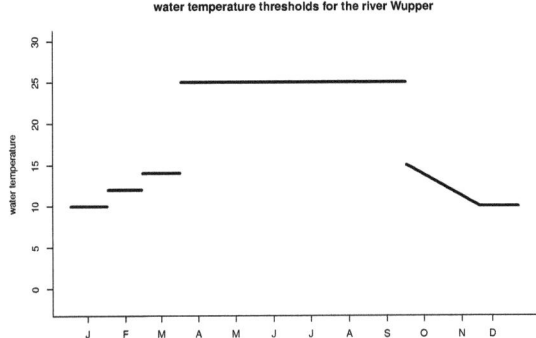

Figure 1.2: Relevant temperature thresholds for the Brown Trout.

determine a standard or reference year and how to measure deviations of the current year. A solution based on landmark-based curve registration is presented in Chapter 3.

Relevant Characteristics of the River Wupper The river Wupper has an overall length of 115km from its origin to its mouth and its drainage area covers 815km^2, see Figure 1.1. The mean annual discharge is about 17m^3/sec near its mouth and approximately 9m^3/sec at the upstream measurement site "Laaken". Before the river enters the city of Wuppertal its bed is left almost natural although there are a large number of smaller and bigger dams. Within the city bounds of Wuppertal it is channeled to an average width of about 7m.

1 Introduction

1.2 Outline of this Thesis

Before looking in detail at the applications a review on the theory of the most important statistical tools employed in the later chapters is given in Chapter 2. Here the topics "generalized additive models", "EM-algorithm", "curve registration" and "approximate dynamic factor models" are addressed.

In Chapter 3 an application of landmark-based curve registration to the water temperature data of the river Wupper is described. We define four different landmark criteria that partly exploit the multivariate structure of the data to identify characteristics that can be observed repeatedly in every year thereby giving hints if the current year is running ahead or behind some previously defined reference year. A special challenge in the definition of landmarks is to formulate online criteria, i. e. criteria that only use contemporaneously available data to decide whether a certain landmark has been reached.

Chapter 4 is dedicated to the forecasting of hourly water temperature readings based on the historical water and air temperature. Here approximate dynamic factor models will be employed. As both, water and air temperature, are measured on an hourly basis they can be interpreted as 24-dimensional time series and from both we will extract common factors. The interdependence between both temperatures will then be modelled on the factor level. We examine three different approaches to factor estimation and compare the performance through an out-of-sample forecast with a classical time series approach as well as with univariate water temperature forecasting models from the hydological field because multivariate approaches are lacking in the respective literature.

In Chapter 5 we suggest a general routine for energy demand forecasts based on a two stage modelling. First, deterministic pattern dependent on external calendarial or meteorological covariates are extracted using a generalized additive model for each hour (or half-hour) separately. We impose a dynamic factor structure on the remaining residuals. The methodology is applied not only to the heat demand data provided by the municipal utility of Wuppertal as literature on heat demand forecasting is sparse. Instead we present a second data example where we forecast the aggregate electricity demand of the state Victoria, Australia. For electricity demand there exists a wide literature and

1 Introduction

we compare three different factor models with the quite popular periodic autoregressions.

This thesis is based on the following papers:

Kauermann, G. and Mestekemper, T. (2010). A Short Note on Quantifying and Visualizing Yearly Variation in Online Monitored Temperature Data. *Statistical Modelling* (to appear).

Mestekemper, T., Windmann, M. and Kauermann, G. (2010). Functional Hourly Forecasting of Water Temperature. *International Journal of Forecasting* 26. 684-699.

Mestekemper, T., Kauermann, G. and Smith, M. (2010). A Comparison of Periodic Autoregressive and Dynamic Factor Models in Intraday Energy Demand Forecasting. *International Journal of Forecasting* (submitted).

2 Theoretical Background

In this chapter a review on the theory of the most important methods applied in the following three chapters will be given. It starts with a brief and by far not exhaustive overview of generalized additive models in Section 2.1 and is followed by a short introduction to the expectation maximization (EM) algorithm in Section 2.2. The stress of this thesis is on landmark-based curve registration and on approximate dynamic factor models which will be discussed in more detail in Sections 2.3 and 2.4, respectively.

2.1 Generalized Additive Models

Suppose that some random vector $\boldsymbol{Y} = (Y_1, \ldots, Y_N)^\top$ follows an exponential family distribution, i.e. $\boldsymbol{Y} \sim \exp\left[(\boldsymbol{Y}\theta - b(\theta))/\phi + c(\boldsymbol{Y}, \phi)\right]$ where θ is the canonical parameter, ϕ is the scale or dispersion parameter and $c(\cdot, \cdot)$ is some function (see McCullagh & Nelder, 1999). $\boldsymbol{Y}$ is assumed to depend on p observed covariate vectors $\boldsymbol{X}_j = (X_{j1}, \ldots, X_{jN})^\top, j = 1, \ldots, p$. Let the realisations of these vectors be denoted by $\boldsymbol{y}, \boldsymbol{x}_1, \ldots, \boldsymbol{x}_p$. The most general model that can be formulated is given by

$$g(\mu_i) = f(x_{1i}, \ldots, x_{pi}) + \epsilon_i, \qquad (2.1)$$

where $g(\cdot)$ is an appropriate link function, $\mu_i := \mathrm{E}(Y_i)$ and ϵ_i the residual. f is a possibly smooth but otherwise unspecified function that shall be estimated from the data by applying smoothing techniques. Note that in the literature there seems to be a consensus to call a function "smooth" if it is twice continously differentiable but depending on the basis functions employed the funtional term can also be piecewise linear. In this thesis only spline smoothing (see Ruppert, Wand & Carroll, 2003 or

Wood, 2006 as comprehensive textbook references) will be treated but there are other possibilities such as kernel smoothing, for instance. Unfortunately, model (2.1) has two major drawbacks: Firstly, for $p > 2$ one faces huge numerical obstacles that can impede its estimation, that is, the model suffers from the so-called "curse of dimensionality". Secondly, if the estimation is possible anyhow its interpretation can be very difficult as the impact of a single covariate cannot be extracted easily. Therefore this approach is of little interest in practise. Hastie & Tibshirani (1990) proposed to model the influence of the covariates additively as univariate or, at most, bivariate (smooth) functions. Such a generalized additive model (GAM) can be structured like

$$g(\mu_i) = \boldsymbol{x}_i\boldsymbol{\beta} + f_1(x_{1i}) + f_2(x_{2i}) + f_3(x_{3i}, x_{4i}) + \cdots + \epsilon_i. \tag{2.2}$$

Here $\boldsymbol{x}_i$ is the i-th row of a design matrix $\boldsymbol{X}$. Note that (2.2) suffers from a identifiabilty problem. For example, a constant could be substracted from f_1 and added to f_2 without changing the fit. Hence, indentifiability constraints have to be introduced.

The first term on the right hand side of (2.2) stands for the parametric part of the model and the remaining terms represent the non-parametric part. Models of this type are therefore often called "semiparametric" to emphasize that they form some kind of a compromise between purely parametric models and fully non-parametric ones like kernel smoothers. However, the term "semiparametric" is missleading as the non-parametric part of (2.2) still contains parameters as will be shown later. The purpose of naming it "non-parametric" is to cleary differentiate the "flexible" part of the model from the static part.

Both main disadvantages of (2.1) have been eliminated by using model (2.2). We now face the estimation of several low dimensional functions instead of a single high dimensional one and those functions can be interpreted more easily. In the remainder of this section a detailed description of penalized splines will be given which represent only one possibility to estimate the possibly smooth terms $f.$ in equation (2.2) from the data.

2.1.1 Penalized Splines

A spline of degree m is defined to be piecewisely composed of polynomials of degree m or less. By choosing m the smoothness of the spline, i.e. the number of continuous derivatives is selected. In a more general framework an univariate unknown function $f(\boldsymbol{x})$ (bivariate functions will be treated in Section 2.1.4) can be written as a linear combination of appropriate basis functions $b_k(\cdot)$ by setting

$$f(x_i) = \sum_{k=1}^{K} b_k(x_i) v_k, \qquad (2.3)$$

where the v_k are basis coefficients. A GAM of the structure

$$g(\mu_i) = \boldsymbol{X}_i \boldsymbol{\beta} + f_1(x_{1i}) + f_2(x_{2i}) + \cdots + \epsilon_i, \quad i = 1, \ldots, N,$$

can then be rewritten as generalized linear model (GLM)

$$g(\mu_i) = \boldsymbol{X}_i^* \boldsymbol{\beta}^* + \epsilon_i, \qquad (2.4)$$

by setting

$$\boldsymbol{X}^* = \left[\boldsymbol{X}, b_1^1(\boldsymbol{x}_1), \ldots, b_{K_1}^1(\boldsymbol{x}_1), b_1^2(\boldsymbol{x}_2), \ldots, b_{K_2}^2(\boldsymbol{x}_2), \ldots \right],$$

and

$$\boldsymbol{\beta}^* = \begin{bmatrix} \boldsymbol{\beta} \\ \boldsymbol{v}^1 \\ \boldsymbol{v}^2 \\ \vdots \end{bmatrix},$$

where $\boldsymbol{v}^m = (v_1^m, \ldots, v_{K_m}^m)^\top$, $m = 1, 2, \ldots$. Model (2.4) can be estimated by applying the Fisher-Scoring algorithm which is implemented in many standard software packages.

Now, as mentionend above, a spline is a special function that is constructed piecewisely by "glueing" together polynomials. The breakpoints where two parts of the function are merged are called "knots". For a given set of knots $\kappa_1, \ldots, \kappa_K$ a cubic spline can be build from a truncated polynomial basis (see Section 2.1.4 for alternative bases), i.e.

$$f(x) = \beta_0 + x\beta_1 + x^2 \beta_2 + x^3 \beta_3 + \sum_{k=1}^{K} (x - \kappa_k)_+^3 u_k. \qquad (2.5)$$

2 Theoretical Background

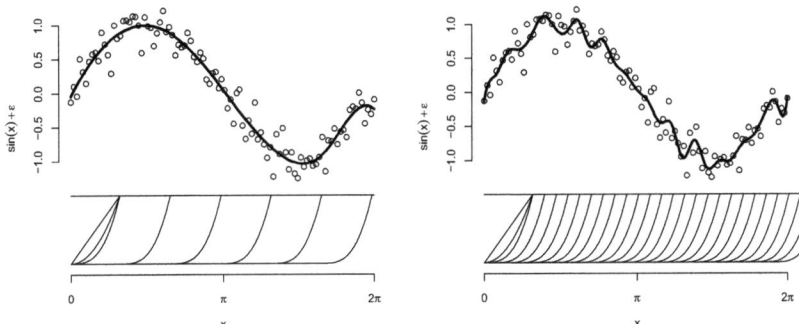

Figure 2.1: A cubic truncated polynomial spline fitted to sinusoidal data with gaussian errors. In the left panel five equidistant knots were used, in the right panel 25. The basis functions are sketched in the lower part of both plots.

Note that the basis coefficients, formerly annotated as $v.$, have been divided into two groups by using different annotations: $\beta.$ and $u.$. The advantage of this distiction will be clarified in Section 2.1.2. The subscipt $(z)_+$ denotes the positive part of the function z, i. e.

$$(z)_+ = \begin{cases} z, & z \geq 0 \\ 0, & z < 0. \end{cases}$$

Figure 2.1 shows a cubic spline fitted to the values of a sinusoidal function which were observed with guassian errors. The left panel depicts a setting where five equidistant knots were used and the result seems to describe the data well except for an edge effect on the right side of the plot. When the number of knots is increased to 25, see the right panel, the result is a smooth but "wiggly" fit. In this case the data were "overfitted". This demonstrates that the choice of the number (and the location) of the knots has a strong influence on the fit. However, in order to achieve an automatic smoothing procedure Ruppert, Wand & Carroll (2003) suggest, for example, to choose the number of knots by the formula

$$K = \min\left(\frac{1}{4} \times \text{number of unique covariate values}, 35\right).$$

Their idea is to choose a sufficient number so that a good fit can be achieved without allowing the set of knots to grow excessively for huge datasets. As knot location they propose samples quantiles:

$$\kappa_k = \left(\frac{k}{K+1}\right)\text{-th sample quantile of the unique covariate values.}$$

This guarantees that the model is flexible enough where a lot of information is available while only few knots are located where data are sparse. Having fixed the number and location of knots by these or similar formulas a wiggly fit can be avoided by introducing a roughness penalty for each of the functional terms $f_m, m = 1, 2, \ldots$ that is given by

$$\begin{aligned}\lambda_m J(f_m) &= \lambda_m \int [f_m''(z)]^2 \mathrm{d}z \\ &= \lambda_m \sum_{k,l=1}^{K_j} v_k^m v_l^m \int b_k^{m''}(z) b_l^{m''}(z) \mathrm{d}z =: \lambda_m (\boldsymbol{v}^m)^\top \boldsymbol{S}_m \boldsymbol{v}^m.\end{aligned} \quad (2.6)$$

Here λ_m is called a smoothing parameter and $J(f_m)$ is a roughness measure for function f_m. Equation (2.6) demonstrates that the latter can be written as quadratic form where $\boldsymbol{S}_m$ is a positiv semi-definite matrix of known coefficients. For given smoothing parameters $\lambda_m, m = 1, 2, \ldots$ and given starting values $\boldsymbol{\mu}^{(0)}$ and $\boldsymbol{\beta}^{*(0)}$ the estimator $\hat{\boldsymbol{\beta}}^*$ can be calculated by penalized iteratively re-weighted least squares (P-IRLS, see Wood, 2006):

1. Given the current values for $\boldsymbol{\mu}^{(s)}$ and $\boldsymbol{\beta}^{*(s)}$ calculate:

$$w_i \propto \frac{1}{V(\mu_i^{(s)})g'(\mu_i^{(s)})^2} \quad \text{and} \quad z_i = g'(\mu_i^{(s)})(y_i - \mu_i^{(s)}) + \boldsymbol{X}_i^* \boldsymbol{\beta}^{*(s)}, i = 1, \ldots, N,$$

where $V(\mu_i^{(s)}) = \mathrm{Var}(Y_i)/\phi$ is the variance function and ϕ is the dispersion parameter of the corresponding exponential family distribution.

2. Calculate

$$\boldsymbol{\beta}^{*(s+1)} = \underset{\boldsymbol{\beta}^*}{\operatorname{argmin}} ||\sqrt{\operatorname{diag}(\boldsymbol{w})}(\boldsymbol{z} - \boldsymbol{X}^*\boldsymbol{\beta}^*)||^2 + \lambda_1(\boldsymbol{v}^1)^\top \boldsymbol{S}_1 \boldsymbol{v}^1 + \lambda_2(\boldsymbol{v}^2)^\top \boldsymbol{S}_2 \boldsymbol{v}^2 + \ldots,$$

with $\boldsymbol{w} = (w_1, \ldots, w_N)^\top$.

Iterate 1. and 2. until convergence.

Optimal smoothing parameters can be found by using a generalized cross-validation criterion (GCV, see Wood, 2006), for instance. If more than one functional terms are involved, a grid search can be performed to identify the optimal combination of smoothing parameters. This approach will not be explained in detail in this thesis. Here the mixed model approach is mainly pursued which will be described in the following part of this section.

2.1.2 Representing GAMs as Generalized Linear Mixed Model

It will be shown that every functional term with a quadratic penalty as in (2.6) can be integrated in a linear mixed model framework, that is, every GAM (or, more generally speaking, every generalized additive mixed model, GAMM) can be written as generalized linear mixed model (GLMM). This is desirable as mixed model software that is readily available in many statistical software packages can be used for estimation and, furthermore, the optimal smoothing parameter(s) result more or less automatically from the estimation routine so that a search algorithm via GCV or similar criteria is unnecessary.

GLMMs are of interest in many scientific fields. By incorporating random effects differences between individuals in a longitudinal study can be modelled, for instance. However, as will be demonstrated the random component is also helpful when estimating spline models. Let $\boldsymbol{\mu}^u := \operatorname{E}(\boldsymbol{y}|\boldsymbol{u})$ where $y_i|\boldsymbol{u}$ follows an exponential family distribution. A GLMM is defined to be a model of the structure

$$g(\mu_i^u) = \boldsymbol{x}_i \boldsymbol{\beta} + \boldsymbol{z}_i \boldsymbol{u} + \epsilon_i \quad \text{with} \quad \boldsymbol{u} \sim \operatorname{N}(\boldsymbol{0}, \boldsymbol{\Sigma}_\varphi^u),$$

where $\boldsymbol{z}_i$ is the i-th row of the model matrix $\boldsymbol{Z}$ for the random effects $\boldsymbol{u}$ and $\boldsymbol{\Sigma}_\varphi^u$ is the corresponding covariance matrix of the random effects that depends on some distributional parameters $\boldsymbol{\varphi}$. Assume that a GAMM with univariate functional terms (the

2 Theoretical Background

bivariate case will be treated in Section 2.1.4) of the form

$$g(\mu_i^u) = \boldsymbol{x}_i\boldsymbol{\beta} + f_1(x_{1i}) + f_2(x_{2i}) + \cdots + \boldsymbol{z}_i\boldsymbol{u} + \epsilon_i, \tag{2.7}$$

shall be written as GLMM. For simplicity it will be demonstrated by means of a single functional term

$$f(\boldsymbol{x}) = \sum_{k=1}^{K} b_k(\boldsymbol{x})v_k =: \boldsymbol{X}^f\boldsymbol{v},$$

with associated roughness measure $J(f) = \boldsymbol{v}^\top \boldsymbol{S}\boldsymbol{v}$ as in (2.6) that this is possible. Following the Bayesian argumentation in Wood (2006) the assumption that f is rather smooth than wiggly can be formalized by choosing an exponential distribution with rate parameter λ as prior for $\boldsymbol{v}$, i.e.

$$f_v(\boldsymbol{v}) \propto \exp\left(-\lambda \frac{\boldsymbol{v}^\top \boldsymbol{S}\boldsymbol{v}}{2}\right). \tag{2.8}$$

In general $\boldsymbol{S}$ is not of full rank, that is, the prior f_v is generally improper. Let $\boldsymbol{S} =: \boldsymbol{U}\boldsymbol{D}\boldsymbol{U}^\top$ denote the eigendecomposition of $\boldsymbol{S}$ so that $\boldsymbol{D}$ is a diagonal matrix of the eigenvalues arranged in decreasing order and the columns of $\boldsymbol{U}$ are the corresponding eigenvectors. Remember that $\boldsymbol{S}$ is positiv semidefinite and define $\boldsymbol{D}_+$ as the largest submatrix of $\boldsymbol{D}$ with strictly positive values on the main diagonal. After reparameterising $\boldsymbol{v}_u := \boldsymbol{U}\boldsymbol{v}$ the new parameter vector can be split into $\boldsymbol{v}_u =: (\boldsymbol{v}_p^\top, \boldsymbol{v}_{up}^\top)^\top$ where $\boldsymbol{v}_p$ denotes the penalized coefficients corresponding to the eigenvalues in $\boldsymbol{D}_+$ and $\boldsymbol{v}_{up}$ are the remaining unpenalized coefficients. By setting $1/\lambda = \sigma^2$ the prior in (2.8) can be written as

$$f_v(\boldsymbol{v}) \propto \exp\left(-\frac{\boldsymbol{v}_u^\top \boldsymbol{D}\boldsymbol{v}_u}{2\sigma^2}\right) = \exp\left(-\frac{\boldsymbol{v}_p^\top \boldsymbol{D}_+\boldsymbol{v}_p}{2\sigma^2}\right).$$

That is, the prior for the penalized parameters $\boldsymbol{v}_p$ is multivariate normal while it is completely uninformative for the remaining coefficients $\boldsymbol{v}_{up}$. This fact fits greatly into a mixed model framework. If $\boldsymbol{U} =: [\boldsymbol{U}_p, \boldsymbol{U}_{up}]$ is divided into two parts corresponding to penalized and unpenalized parameters, respectively, a mixed model can be formulated by setting $\boldsymbol{X}_{up} := \boldsymbol{X}^f\boldsymbol{U}_{up}$ and $\boldsymbol{X}_p := \boldsymbol{X}^f\boldsymbol{U}_p$:

$$\boldsymbol{X}_{up}\boldsymbol{v}_{up} + \boldsymbol{X}_p\boldsymbol{v}_p \quad \text{with} \quad \boldsymbol{v}_p \sim \mathrm{N}(\boldsymbol{0}, \boldsymbol{D}_+^{-1}/\lambda).$$

2 Theoretical Background

To obtain the familiar mixed model annotation some last reparametrisations are necessary. Define $\boldsymbol{\beta}_{\text{up}} := \boldsymbol{v}_{\text{up}}$, $\boldsymbol{u}_{\text{p}} := \sqrt{\boldsymbol{D}_+}\boldsymbol{v}_{\text{p}}$ and $\boldsymbol{Z}_{\text{p}} := \boldsymbol{X}_{\text{p}}\sqrt{\boldsymbol{D}_+^{-1}}$ yields

$$\boldsymbol{X}_{\text{up}}\boldsymbol{\beta}_{\text{up}} + \boldsymbol{Z}_{\text{p}}\boldsymbol{u}_{\text{p}} \quad \text{with} \quad \boldsymbol{u}_{\text{p}} \sim \text{N}(\boldsymbol{0}, \boldsymbol{I}/\lambda).$$

Here the smoothing parameter λ is a fixed parameter that can be estimated along with the fixed effects parameter vector $\boldsymbol{\beta}_{\text{up}}$. Now the integration of a smooth term in a GLMM is straight forward. It only remains to append the columns of $\boldsymbol{X}_{\text{up}}$ to the already existing fixed effects matrix $\boldsymbol{X}$ in (2.7) and to combine the random effect matrices $\boldsymbol{Z}$ and $\boldsymbol{Z}_{\text{p}}$ in an analogous way. Simultaneously, the parameter vectors are merged, i.e. $\boldsymbol{\beta}_{\text{new}} := (\boldsymbol{\beta}^\top, \boldsymbol{\beta}_{\text{up}}^\top)^\top$ and $\boldsymbol{u}_{\text{new}} := (\boldsymbol{u}^\top, \boldsymbol{u}_{\text{p}}^\top)^\top$. The same procedure is repeated for all smooth terms in (2.7).

2.1.3 Estimation

After having demonstrated that every GAMM can be written as GLMM it remains to present an estimation routine for the latter class of models. Looking at a GLMM in the general form

$$g(\mu_i^u) = \boldsymbol{X}_i\boldsymbol{\beta} + \boldsymbol{Z}_i\boldsymbol{u} + \epsilon_i, \quad \boldsymbol{u} \sim \text{N}(\boldsymbol{0}, \boldsymbol{\Sigma}_\varphi^u), \quad \boldsymbol{\mu}^u := \text{E}(\boldsymbol{y}|\boldsymbol{u}),$$

where $y_i|\boldsymbol{u}$ follows a distribution from the exponential family with associated link function g the parameters to be estimated are the fixed effects $\boldsymbol{\beta}$ and the variance components of the random effects $\boldsymbol{\varphi}$. This is done by maximizing the (log-)likelihood of the joint distribution

$$f_{\boldsymbol{\beta},\boldsymbol{\varphi}}(\boldsymbol{y}, \boldsymbol{u}) \propto |\boldsymbol{\Sigma}_\varphi^u|^{-\frac{1}{2}} \exp\left(\log f(\boldsymbol{y}|\boldsymbol{u}) - \frac{1}{2}\boldsymbol{u}^\top(\boldsymbol{\Sigma}_\varphi^u)^{-1}\boldsymbol{u}\right),$$

where $|\cdot|$ is the determinant. A likelihood function that focuses on the parameters of interest can be obtained by integrating out the random effects and using the observed response $\boldsymbol{y}_{\text{obs}}$:

$$L(\boldsymbol{\beta}, \boldsymbol{\varphi}) \propto |\boldsymbol{\Sigma}_\varphi^u|^{-\frac{1}{2}} \int \exp\left(\log f(\boldsymbol{y}_{\text{obs}}|\boldsymbol{u}) - \frac{1}{2}\boldsymbol{u}^\top(\boldsymbol{\Sigma}_\varphi^u)^{-1}\boldsymbol{u}\right) \mathrm{d}\boldsymbol{u}. \quad (2.9)$$

2 Theoretical Background

Calculation of the log-likelihood term $l(\boldsymbol{\beta}, \boldsymbol{u}) = \log f(\boldsymbol{y}_{\text{obs}}|\boldsymbol{u})$ in (2.9) is numerically difficult for higher dimensions of $\boldsymbol{u}$ but can be replaced by its Laplace approximation what leads to the approximate likelihood function of the joint distribution

$$L^*(\boldsymbol{\beta}, \boldsymbol{\varphi}) \propto |\boldsymbol{\Sigma}^u_{\boldsymbol{\varphi}}|^{-\frac{1}{2}} \int \exp\left(-\frac{1}{2\phi}\left\|\boldsymbol{W}^{-\frac{1}{2}}(\boldsymbol{z} - \boldsymbol{X}\boldsymbol{\beta} - \boldsymbol{Z}\boldsymbol{u})\right\|^2 - \frac{1}{2}\boldsymbol{u}^\top (\boldsymbol{\Sigma}^u_{\boldsymbol{\varphi}})^{-1}\boldsymbol{u}\right) \mathrm{d}\boldsymbol{u}, \tag{2.10}$$

where $\boldsymbol{W}$ is a diagonal matrix with

$$W_{ii} = \frac{1}{V(\mu_i^b) g'(\mu_i^b)^2} \quad \text{and} \quad z_i = g'(\mu_i^b)(y_i - \mu_i^b) + \boldsymbol{X}_i \boldsymbol{\beta} + \boldsymbol{Z}_i \boldsymbol{u}.$$

Equation (2.10) is also called the integrated quasi-likelihood function and the approximate estimation algorithm described below is known as penalized quasi-likelihood (PQL). PQL was suggested in a Bayesian framework by Laird (1978) and the algorithm was justified by Schall (1991) and Breslow & Clayton (1993) who relate the PQL criterion to a Fisher Scoring algorithm developed by Green (1987).

Starting with some initial estimates $\hat{\boldsymbol{\beta}}^{(0)}$ and $\hat{\boldsymbol{u}}^{(0)}$ the fixed effects and the variance components can be computed by iterating the following steps to convergence:

1. Given $\hat{\boldsymbol{\beta}}^{(s)}$ and $\hat{\boldsymbol{u}}^{(s)}$ calculate $\hat{\boldsymbol{\mu}}^{u(s)} = g^{-1}(\boldsymbol{X}\hat{\boldsymbol{\beta}}^{(s)} + \boldsymbol{Z}\hat{\boldsymbol{u}}^{(s)})$, $\boldsymbol{z}$ and $\boldsymbol{W}$.

2. Estimate the linear mixed model

$$\boldsymbol{z} = \boldsymbol{X}\boldsymbol{\beta} + \boldsymbol{Z}\boldsymbol{u} + \boldsymbol{\epsilon}, \quad \boldsymbol{u} \sim \mathrm{N}(\boldsymbol{0}, \boldsymbol{\Sigma}^u_{\boldsymbol{\varphi}}), \quad \boldsymbol{\epsilon} \sim \mathrm{N}(\boldsymbol{0}, \boldsymbol{W}^{-1}\phi), \tag{2.11}$$

to obtain $\hat{\boldsymbol{\beta}}^{(s+1)}$, $\hat{\boldsymbol{u}}^{(s+1)}$, $\hat{\boldsymbol{\varphi}}^{(s+1)}$ and $\hat{\phi}^{(s+1)}$.

The (ordinary) linear mixed model (2.11) can be fitted by maximising the corresponding profile likelihood that is given by

$$L_{\mathrm{p}}(\boldsymbol{\varphi}) = \frac{1}{\sqrt{(2\pi\hat{\sigma}^2_{\boldsymbol{\varphi}})^N |\boldsymbol{\Sigma}_{\boldsymbol{\varphi}}|}} \exp\left(-\frac{1}{2\hat{\sigma}^2_{\boldsymbol{\varphi}}}\left(\boldsymbol{y} - \boldsymbol{X}\hat{\boldsymbol{\beta}}_{\boldsymbol{\varphi}}\right)^\top \boldsymbol{\Sigma}^{-1}_{\boldsymbol{\varphi}}\left(\boldsymbol{y} - \boldsymbol{X}\hat{\boldsymbol{\beta}}_{\boldsymbol{\varphi}}\right)\right),$$

with respect to $\boldsymbol{\varphi}$. Here $\hat{\boldsymbol{\beta}}_{\boldsymbol{\varphi}}$ and $\hat{\sigma}^2_{\boldsymbol{\varphi}}$ are standard estimators written as functions of the variance components $\boldsymbol{\varphi}$. The maximisation of $L_{\mathrm{p}}(\boldsymbol{\varphi})$ is easy as a multivariate version of the Newton-Raphson algorithm can be applied.

2.1.4 Spline Bases and Penalties

A possibly smooth term f can be represented as a linear combination of a certain set of basis functions $b_k, k = 1, \ldots, K$ as was shown in equation (2.3). Deliberately, in this section about generalized additive models the functions $b_k(\cdot)$ have remained unspecified as there is a huge amount of possible bases a statistician can choose from depending on his data and the purpose of his analysis. In Section 2.1.1 a truncated polynomial basis was introduced to motivate the need of a penalisation. However, for a large number of knots this basis turns out to be numerically unstable for small values of the smoothing parameter λ. It should also be noted that it can be advantegeous to choose another roughness measure than that given in equation (2.6) while retaining a quadratic form benefits the implementation. In the following three types of basis functions will be briefly presented that were used in the applications of this thesis.

B-Spline Basis The main advantage of B-spline basis functions is that they are only different from zero over a bounded interval, i. e. a B-spline basis function of degree $m+1$ is only positive between $m+3$ knots. This leads to enhanced numerical stability if the set of breakpoints is large. Let $\kappa_1 < \kappa_2 < \ldots < \kappa_{K+m+1}$ be such an ordered set of knots. Then B-spline basis functions can be defined recursively (see de Boor, 1978) by setting

$$B_k^m(x) := \frac{x - \kappa_k}{\kappa_{k+m+1} - \kappa_k} B_k^{m-1}(x) + \frac{\kappa_{k+m+2} - x}{\kappa_{k+m+2} - \kappa_{k+1}} B_{k+1}^{m-1}(x), \quad k = 1, \ldots, K,$$

with

$$B_k^{-1}(x) := \begin{cases} 1, & \kappa_k \leq x < \kappa_{k+1} \\ 0, & \text{else.} \end{cases}$$

A B-spline of order $m+1$ written in the style of equation (2.3) is given by

$$f(x) = \sum_{k=1}^{K} B_k^m(x) v_k.$$

As associated roughness measure Eilers & Marx (1996) propose a difference penalty as approximation of the second squared derivative of f. If first-order differences shall be used this results in

$$J(f) = \sum_{k=1}^{K-1}(v_{k+1} - v_k)^2 = \boldsymbol{v}^\top \begin{bmatrix} 1 & -1 & 0 & 0 & 0 & \cdots \\ 1 & -2 & 1 & 0 & 0 & \cdots \\ 0 & 1 & -2 & 1 & 0 & \cdots \\ \vdots & \vdots & \ddots & \ddots & \ddots & \ddots \end{bmatrix} \boldsymbol{v}.$$

Higher order differences that lead to smoother fits can be implemented analogously.

Thin Plate Spline Basis Thin plate splines were suggested by Duchon (1977) and have been given their name as the roughness measure is proportional to the bending energy that would result if a thin plate was deformed in the same shape as the functional term f. A major advantage of using a thin plate spline basis is that it theoretically permits to estimate a high dimensional function, that is f can be a function of d covariates. For $2m > d$ the roughness penalty is defined by

$$J_{m,d}(f) = \int_\mathbb{R} \cdots \int_\mathbb{R} \sum_{\nu_1 + \cdots + \nu_d = m} \frac{m!}{\nu_1! \cdots \nu_d!} \left(\frac{\partial^m f}{\partial x_1^{\nu_1} \cdots \partial x_d^{\nu_d}} \right)^2 \mathrm{d}x_1 \ldots \mathrm{d}x_d,$$

(see Wood, 2006). A thin plate spline can then be written as

$$f(\boldsymbol{x}) = \sum_{j=1}^{M} \psi_j(\boldsymbol{x})\beta_j + \sum_{i=1}^{N} \eta_{m,d}(||\boldsymbol{x} - \boldsymbol{x}_i||)u_i, \quad \text{with} \quad \boldsymbol{T}^\top \boldsymbol{u} = 0,$$

where $T_{ij} = \psi_j(\boldsymbol{x}_i)$ and $M = \binom{m+d-1}{d}$. Here the functions $\psi(\cdot)$ are those polynomials that span the space of polynomials of degree less than m in $\mathbb{R}^d$ and for that $J_{m,d}$ is zero, i.e. in the mixed model context these functions remain unpenalized and will be categorised as fixed effects. The remaining functions to be penalized are defined by

$$\eta_{m,d}(r) = \begin{cases} \frac{(-1)^{m+1+d/2}}{2^{2m-1}\pi^{d/2}(m-1)!(m-d/2)!} r^{2m-d} \log(r), & \text{for } d \text{ even}, \\ \frac{\Gamma(d/2-m)}{2^{2m}\pi^{d/2}(m-1)!} r^{2m-d}, & \text{for } d \text{ odd}. \end{cases}$$

Further advantages of thin plate splines are that no knots have to be specified as these are given by the observations itselves which means that a certain amount of subjectivity is removed from the estimation process. In addition the estimated function is invariant under rotation and translation of the underlying coordinate system. On the other side this basis comes at a high computational cost and in practice has to be approximated by so-called thin plate regression splines (Wood, 2006). For further details on this topic see Wahba (1990) and Green & Silverman (1994).

Tensor Product Basis Besides the thin plate spline approach a bi- or multivariate spline basis can be defined by a tensor product basis. In the two dimensional case assume that for two covariates x_1 and x_2 arbitrary bases $\mathcal{B} = \{b_k(x_1)|k = 1,\ldots,K_1,\ x_1 \in \mathbb{R}\}$ and $\mathcal{C} = \{c_l(x_2)|l = 1,\ldots,K_2,\ x_2 \in \mathbb{R}\}$ are at hand. A bivariate spline basis can be constructed by taking the tensor product $\mathcal{B} \otimes \mathcal{C}$, i.e.

$$f(\boldsymbol{x}_1, \boldsymbol{x}_2) = \sum_{k=1}^{K_1}\sum_{l=1}^{K_2} b_k(\boldsymbol{x}_1)c_l(\boldsymbol{x}_2)v_{kl}.$$

Let $J_{x_1}(f_{x_1|x_2})$ be the one dimensional penalty assiciated to $\mathcal{B}$ for a given value of x_2 and define $J_{x_2}(f_{x_2|x_1})$ analogously. Then a valid roughness measure can be obtained by integrating out the fixed covariates and adding both terms:

$$J(f_{x_1x_2}, \lambda_{x_1}, \lambda_{x_2}) = \lambda_{x_1}\int_{\mathbb{R}} J_{x_1}(f_{x_1|x_2})\mathrm{d}x_2 + \lambda_{x_2}\int_{\mathbb{R}} J_{x_2}(f_{x_2|x_1})\mathrm{d}x_1$$

where the λ. are to be understood as smoothing parameters in their corresponding direction. Bases for higher dimensions can be constructed in complete analogy. For details concerning the implementation of the penalty term, see Wood (2006).

2.1.5 Computational Issues

In R the estimation of generalized additive models can be performed using the package mgcv created by Simon N. Wood. Wood (2004, 2006, 2008) give a description of the capability of this package. The main function gam() plays the role of a wrapper function that translates the GAM formula and the data passed as arguments into a mixed model and then calls the PQL estimation routine glmmPQL() from the MASS package or in case that no generalized response is involved the function lme() from the nmle package (see Pinheiro & Bates, 2002).

2.2 The EM-Algorithm

The EM algorithm is a possible method to get reliable maximum-likelihood based parameter estimates when a part of the data is missing or unobserved. In many applications it also can be helpful to fomulate an estimation problem with complete data as an incomplete one and to apply the EM algorithm as it requires less computational resources than a Fisher Scoring algorithm, for instance, although it can be significantly slower. Many authors have applied specialized algorithms which are very similar to the EM algorithm before Dempster, Laird & Rubin (1977) formulated a global defintion which today is commonly known. Each iteration of the algorithm consists of the following two steps:

1. **E-Step:** In the *Estimation Step* of iteration s the set of unknown parameters or observations θ is replaced by its estimation from the preceding iteration $\theta^{(s-1)}$ or an initial value $\theta^{(0)}$ in case of the first iteration, respectively. Then a function which is often called the "*Q*-function", that is, the conditional expected value of the log-likelihood given the observed data and the current estimate of the unobserved values is evaluated:
$$Q(\theta, \theta^{(s-1)}) = \mathrm{E}_{\theta^{(s-1)}}(\log L(\theta)|\boldsymbol{y}),$$
where $\boldsymbol{y}$ are the observed (incomplete) data and $L(\cdot)$ is the likelihood function.

2. **M-Step:** In the *Maximization Step* of iteration s the Q-function is maximized and a new estimator $\theta^{(s)}$ is chosen:
$$\theta^{(s)} \in \{\theta \,|\, \theta = \operatorname*{argmax}_{\theta} Q(\theta, \theta^{(s-1)})\}.$$

Both steps are repeated until either $|Q(\theta, \theta^{(s)}) - Q(\theta, \theta^{(s-1)})|$ or $|\theta^{(s)} - \theta^{(s-1)}|$ is sufficiently small. A detailed description of the EM algorithm can be found in Dempster, Laird & Rubin (1977), Tanner (1992) or in McLachlan & Krishnan (1997) where the latter is currently the only textbook which is fully dedicated to this algorithm and which contains a great number of extensions and examples.

2.3 Curve Registration

Functional data analysis (FDA) is a term coined by Ramsay & Silverman (2005) and recently further discussed in Ferraty & Vieu (2006) and Ramsay, Hooker & Graves (2009). The field is concerned with the analysis of random functional observations or discrete values that can be considered to be discretisized versions of functional observations. Without loss of generality these functional observations can be considered as functions in time t. With the technological advances since the last decade of the 20th century FDA has gained importance as the outcome of more and more experiments are functional. Often these functional observations exhibit a common pattern or similar features like extreme points, for example. However, due to the functional form these features may not only differ by their functional value, i.e. in amplitude, in addition they may be shifted in time, i.e. in phase. Curve registration is a technique from the field of FDA that tackles the latter aspect and is apt to align curves pairwise or towards some reference curve depending on the methodology. The first part of this section gives a brief overview of the literature on curve registration. Section 2.3.2 concentrates on landmark-based time-warping, a special version of curve registration that will be used with some modifications in Chapter 3. Section 2.3.3 treats monotone smoothing with quadratic programming. Although a topic of its own, monotone smoothing is presented within this framework as it forms a desirable improvement of curve registration.

2.3.1 History and Recent Advances

An early example for curve registration is Sakoe & Chiba (1978) who suggest a dynamic programming method for the elimination of speaking rate variation in order to enable the automated recognition of spoken words. Another interesting aspect when looking at a sample of functional observations is the definition of a meaningful average function. When the cross-sectional mean over all observations is taken, features that are clearly exposed in each single curve may be smeared or even disappear. A typical example for this kind of problem is presented in Ramsay, Bock & Gasser (1995). They provide a set of (smoothed) height acceleration curves of children observed over the period from an age of four to adulthood. Every child in the sample experiences a pubertal growth spurt that results in a peak in the acceleration curves followed by a global minimum when the growth spurt terminates. Although these features are clearly exhibited in each sample curve, the cross-sectional mean yields unsatisfying results, meaning that both global extreme points are less exposed as in any of the sample curves. To tackle this problem Kneip & Gasser (1988, 1992) suggest to compute a *structural mean*, that is, to align the curves prior to compute a cross-sectional mean function. Given a set of curves $\{x_i(t) | x_i : [0, T] \to \mathbb{R}, i = 1, \ldots, I\}$ they presume the general model

$$\mu(t) = \alpha_i(t) x_i \left(\pi_i(t) \right) + \epsilon_i(t), \qquad (2.12)$$

where $\mu(t)$ is the structural mean function, $\alpha_i(t)$ is a non-uniform amplitude modulation function, $\pi_i(t)$ is a non-uniform strictly monotonically increasing time transformation, $x_i(t)$ is the i-th possibly smooth sample curve that was observed with noise function $\epsilon_i(t)$ with $\mu, \alpha_i, \epsilon_i : [0, T] \to \mathbb{R}$ and $\pi_i : [0, T] \to [0, T]$. They focus on estimating the time transformation or time-warping function $\pi_i(t)$ by identifying so-called *structural functionals* or landmarks, i.e. characteristics that can be found in every sample curve. As structural functionals they propose to take local extreme points, inflection points and points where the slope of $x_i(t)$ crosses certain thresholds. Having found the locations of these structural functionals (these are called structural points) in every sample curve, their average locations can be computed. Given these informations the time-warping functions $\pi_i(t)$ can be built by either linear interpolating or smoothing the two-dimensional dataset of average and observed time points. The procedure will be described in detail in Section 2.3.2. Applications can be found in Gasser et al. (1990, 1991a, 1991b) and Gasser & Kneip (1995) address some practical aspects of the estimation routine.

Although landmark-based curve registration is considered as benchmark or even best working methodology in many publications such as Ramsay & Li (1998) or Gervini & Gasser (2004, 2005), it is inadequate for large curve samples. Often an automatic calculation of the structural points is not possible and they have to be set by hand. Later research therefore focuses on developing automatic procedures for the calculation

2 Theoretical Background

of time-warping functions. Wang & Gasser (1997, 1998, 1999) present an approach based on dynamic programming. However, it is only apt for pairwise alignment of two curves so that the knowledge of a master curve is necessary if a large sample is to be registered. Ramsay & Li (1998) extend the work of Silverman (1995) and suggest a continous fitting criterion for the model

$$\mu(t) = x_i\left(\pi_i(t)\right) + \epsilon_i(t), \tag{2.13}$$

which is similar to (2.12) but does not allow the curves to vary in amplitude. This leads to an overfitting problem when too much amplitude variation is present in the sample curves. Rønn (2001) presents a maximum likelihood routine for shift registration, that is, the curves may differ in amplitude but the time-warping functions $\pi_i(t)$ are restricted to have a slope equal to one. Gervini & Gasser (2004, 2005) propose another approach to self-modelling warping functions that is motivated by landmark-based curve registration. They extend model (2.13) in the sense that they allow for a uniform amplitude modulation in the style of

$$\mu(t) = \alpha_i x_i\left(\pi_i(t)\right) + \epsilon_i(t). \tag{2.14}$$

Similar to the methodology presented in Ramsay & Li (1998) their method suffers from an overfitting problem if this assumption is violated. An alternative technique using the continous wavelet transform to estimate the time-warping functions was proposed by Bigot (2006). A Bayesian approach to curve registration was presented by Telesca & Inoue (2008) who consider the model

$$\mu(t) = \gamma_i + \alpha_i x_i\left(\pi_i(t)\right) + \epsilon_i(t), \tag{2.15}$$

that is, in addition to (2.14) they allow the curves to have a varying level. Recently, Liu & Yang (2009) described a procedure for simultaneous registration and clustering of curves.

Landmark-based curve registration in two and three dimensions also refered to as image registration has received much attention in medicine and neuro biology. However, the development in this field seems to be more or less independent of the literature presented above and shall therefore not be discussed in depth in this thesis. An extensive literature review can be found in Zitová & Flusser (2003), for example.

The landmark-based approach to curve registration as suggested by Kneip & Gasser (1992) will now be described in detail as it will be applied in Chapter 3.

2.3.2 Landmark-based Curve Registration

As pointed out in the preceding section Kneip & Gasser (1992) use the most general model for time-warping. Its only limitations lie in the restricted applicability for large curve samples when the automatic recognition of landmarks is impossible. However, this is not the case in the application example presented in Chapter 3. Furthermore, the data at hand are time series data and a methodology shall be developed that allows application in an "online" style, i.e. it shall be possible to incorporate new measurements of the time series of interest. In this context the automatic curve registration methods developed later do not work as they require a set of complete functional observations. This section therefore focuses on landmark-based curve registration as proposed by Kneip & Gasser (1988, 1992) and Gasser & Kneip (1995).

2.3.2.1 The Definition of Landmarks

Let J denote the time interval of interest, let $C^\nu(J)$ be the set of all ν times continuously differentiable functions on J and let $||v||_J^{(\nu)} := \sum_{s=0}^{\nu} \sup_{t \in J} |v^{(s)}(t)|, v \in C^\nu(J)$ define a norm on $C^\nu(J)$ where $v^{(s)}(t)$ is the s-th derivative of $v(t)$. Without loss of generality we can set $J := [0, T]$.

Kneip & Gasser (1992) assume that a functional observation consists of a smooth signal that is observed with noise, i.e. a typical sample consists of a set of regression curves $\{x_i(t) | x_i : J \to \mathbb{R}, x_i \in C^\nu(J), i = 1, \ldots, I\}$ for some $\nu \geq 2$. The shape of these sample curves is thereby characterized by structural features like local extreme points together with their corresponding amplitude and inflection points. These characteristics shall be captured by defining continuous landmark functionals $L : C^2(J) \to]0, T[\cup \{a\}$ where L takes the argument value of the feature of interest or $L(v) = a \notin J, v \in C^2(J)$

2 Theoretical Background

if it is missing in v. For a local maximum on a subinterval $[t_1, t_2] \subseteq J$ this boils down to

$$L(v) := \begin{cases} \arg\sup\limits_{t \in [t_1, t_2]} v(t), & \text{if } v \text{ possesses a unique supremum in } [t_1, t_2], \\ a, & \text{otherwise.} \end{cases}$$

From the continuity of L follows that $\forall \epsilon > 0 \; \exists \delta > 0$ such that $\forall v, w \in C^2(J)$ with $||v - w||_J^{(2)} < \delta$ follows $|L(v) - L(w)| < \epsilon$.

Kneip & Gasser (1992) suggest the use of three types of landmarks whose definition are listed below. Let thererfore $\mathcal{D}_L \subseteq C^2(J)$ be the subset of all functions $v \in C^2(J)$ with $L(v) \neq a$.

- A local **extreme point** is a functional $L : C^2(J) \to]0, T[\cup \{a\}$ for some $a \notin J$ if the following assumptions hold:

  1. L is continous on $\mathcal{D}_L$,
  2. $\forall v, w \in \mathcal{D}_L$, $v(L(v))$ and $w(L(w))$ are either both local maxima or local minima of v and w, and $\text{sign}(v''(L(v))) = \text{sign}(w''(L(w))) \neq 0$.

- A local **percentage point** is a location where a certain local percentage of total increase or decrease is reached, that is, a functional $L : C^2(J) \to]0, T[\cup \{a\}$ for some $a \notin J$ with $p \in]0, 1[$ with

  1. L is continous on $\mathcal{D}_L$,
  2. $\forall v \in \mathcal{D}_L$ holds $v(L(v)) = pv(\psi_{0,v}) + (1-p)v(\psi_{1,v})$, where $\psi_{0,v}$ and $\psi_{1,v}$ are successive local extreme points of v, v is strictly monotone on the interval $[\psi_{0,v}, \psi_{1,v}]$ and $\psi_{v,0}$ is either a local maximum or local minimum $\forall v \in \mathcal{D}_L$.

- Further characteristics like **inflection points** are captured by the more general definition of an extreme point functional on the derivates of the functions of interest, i.e. for some $u \in \mathbb{N}$ a functional $L : C^{u+2}(J) \to]0, T[\cup \{a\}$ for some $a \notin J$ is called a landmark of order $u+1$ if there exists an extreme point functional $L_u : C^2(J) \to]0, T[\cup \{a\}$ with $L(v) = L_u(v^{(u)}), \forall v \in C^{u+2}(J)$.

Additionally, Kneip & Gasser (1992) postulate consistency among landmarks of the same order. That is, for any two landmark functionals $L, L^* : C^{u+2}(J) \to]0,T[\cup\{a\}$ for some $a \notin J$, $u \in \mathbb{N} \cup \{0\}$, exactly one of the following equations hold:

1. $L(v) < L^*(v), \forall v \in \mathcal{D}_L$,

2. $L(v) = L^*(v), \forall v \in \mathcal{D}_L$,

3. $L(v) > L^*(v), \forall v \in \mathcal{D}_L$.

Note that consistency is fulfilled by definition between extreme points and percentage points as well as among percentage points.

2.3.2.2 Building the Time-Warping Functions

Let $\boldsymbol{L}_i = (L_{i,1}, \ldots, L_{i,K})^\top$, $K \in \mathbb{N}$ be the vector of landmarks in the i-th sample curve with $L_{i,k} = L_k(x_i)$ and assume that the condition $L_{i,r} < L_{i,s}$ holds for $r < s$ if $L_{i,r} \neq a$ and $L_{i,s} \neq a$. In addition, let $\mathcal{H}_K$ be the set of all ordered vectors $\boldsymbol{t} = (t_1, \ldots, t_l)^\top \in (J \cup \{a\})^K$ that fulfill $t_r < t_s$ for all $r < s$ if $t_r \neq a$ and $t_s \neq a$.

Kneip & Gasser (1992) formulate the following requirements for appropriate time-warping functions $\pi_i, i = 1, \ldots, I$ that correspond to a curve sample $\{x_i(t)|x_i : [0,T] \to \mathbb{R}, i = 1, \ldots, I\}$:

1. As no other information is available, for all $i, j \in \{1, \ldots, I\}$ the differences between π_i and π_j shall only depend on differences between the landmarks $\boldsymbol{L}_i$ and $\boldsymbol{L}_j$.

2. For all $k \in \{1, \ldots, K\}$ the landmark $L_{i,k}$ in the i-th curve shall be align to its average location $\bar{L}_k = \frac{1}{I}\sum_{i=1}^{I} L_{i,k}$.

3. For all $i \in \{1, \ldots, I\}$ π_i shall be strictly monotonically increasing.

4. For all $i \in \{1, \ldots, I\}$ π_i shall be a smooth function.

Note that for empirical applications these points may partly be contradictory. For example, the claim of finding smooth strictly monotonically increasing time-warping functions may inhibit the exact alignment of $L_{i,k}$ to its average location $\bar{L}_k$.

2 Theoretical Background

Kneip & Gasser (1992) formulate the general concept of a *shift operator* which is an operator $\Pi : \mathcal{H}_K^2 \to C^1(\mathbb{R})$ that satisfies:

1. $\forall (\boldsymbol{t_0}, \boldsymbol{t_1}) \in \mathcal{H}_K^2$, $\Pi_{(\boldsymbol{t_0},\boldsymbol{t_1})}(\cdot)$ is a strictly monotonically increasing continous real function,

2. $\forall (\boldsymbol{t_0}, \boldsymbol{t_1}) \in \mathcal{H}_K^2 : \Pi_{(\boldsymbol{t_0},\boldsymbol{t_1})}(t_{0,k}) = t_{1,k}$ for all $k = 1, \ldots, K$ with $t_{0,k} \neq a$ and $t_{1,k} \neq a$.

A time-warping function is then defined by

$$\pi_i(\cdot) := \Pi_{(\bar{\boldsymbol{L}},\boldsymbol{L}_i)}(\cdot),$$

where $\bar{\boldsymbol{L}}$ is the vector of the average landmark locations and $\boldsymbol{L}_i$ is the vector of landmarks in the i-th curve.

The simplest way of finding an appropriate shift operator is to add the start and end points of the time interval of interest to the set of landmarks and to define the time warping functions $\pi_i : [0,T] \to [0,T]$ as linear interpolation between adjacent landmarks, i.e.

$$\pi_i(t) = L_{i,k} + (t - \bar{L}_k)\frac{L_{i,k+1} - L_{i,k}}{\bar{L}_{k+1} - \bar{L}_k} \quad \text{for} \quad t \in \left[\bar{L}_k, \bar{L}_{k+1}\right], k = 1, \ldots, K-1.$$

If some landmarks are missing in $\boldsymbol{L}_i$ the technique can be applied to the remaining ones. However, this approach does not provide smooth time-warping functions. In Section 2.3.3 a monotone smoothing technique will be presented that can fix this deficit.

2.3.3 Monotone Smoothing with Quadratic Programming

For monotone smoothing a special property of B-spline bases (see Section 2.1.4) can be exploited. Therefore the strict monotonicity of the the time-warping functions will be relaxed to non-decreasingness for practicability reasons.

A time-warping function $\pi_i(t)$ can be written as linear combination of B-spline basis functions of order $m+1$

$$\pi_i(t) = \sum_{p=1}^{P} B_p^m(t) u_p.$$

2 Theoretical Background

For simplicity assume that $m = 3$ for cubic B-splines and that $B_p^m, p = 1, \ldots, P$ are uniform, non-degenerated B-splines, that is, a grid of equidistant knots is used and degenerated B-splines on the edges of the interval $[0, T]$ are replaced by their non-degenerated versions.

Kelly & Rice (1990) point out that there can be no more sign changes in $\pi_i(t)$ than there are in the sequence of coefficients $\{u_p\}_{p=1,\ldots,P}$. A non-decreasing time-warping function can therefore be found by minimizing the sum of squared residuals

$$Q(\boldsymbol{u}) = \sum_{i=1}^{I} \sum_{k=1}^{K} \left(L_{i,k} - \mathcal{B}(\bar{L}_k)\boldsymbol{u}\right)^2, \qquad (2.16)$$

subject to the constraints that $u_1 \leq u_2 \leq \ldots \leq u_P$, where $\mathcal{B}$ denotes the B-spline basis $\{B_p^m(t) | p = 1, \ldots, P\}$ and $\boldsymbol{u} = (u_1, \ldots, u_P)^\top$. These constraints can be expressed as

$$\boldsymbol{C}\boldsymbol{u} = \begin{pmatrix} -1 & 1 & 0 & 0 & \cdots & 0 \\ 0 & -1 & 1 & 0 & \cdots & 0 \\ \vdots & \ddots & \ddots & \ddots & \ddots & \vdots \\ 0 & 0 & \cdots & 0 & -1 & 1 \end{pmatrix} \boldsymbol{u} \geq 0. \qquad (2.17)$$

The minimization problem (2.16) under the restrictions (2.17) can be solved iteratively by quadratic programming. Let therefore $\boldsymbol{u}^{(s)}$ be the estimate in the s-th iteration step and let $\boldsymbol{u}^{(s+1)} = \boldsymbol{u}^{(s)} + \boldsymbol{\delta}^{(s)}$. A Taylor expansion of (2.16) yields

$$Q\left(\boldsymbol{u}^{(s+1)}\right) = Q\left(\boldsymbol{u}^{(s)}\right) + \frac{\partial Q\left(\boldsymbol{u}^{(s)}\right)}{\partial \boldsymbol{u}} \boldsymbol{\delta}^{(s)} + \left(\boldsymbol{\delta}^{(s)}\right)^\top \frac{\partial^2 Q\left(\boldsymbol{u}^{(s)}\right)}{\partial (\boldsymbol{u})^\top \partial \boldsymbol{u}} \boldsymbol{\delta}^{(s)},$$

so that we have to minimize the quadratic form

$$\left(\boldsymbol{a}^{(s)}\right)^\top \boldsymbol{\delta}^{(s)} + \left(\boldsymbol{\delta}^{(s)}\right)^\top \boldsymbol{B} \, \boldsymbol{\delta}^{(s)}, \qquad (2.18)$$

subject to (2.17) where

$$\boldsymbol{a}^{(s)} = \frac{\partial Q\left(\boldsymbol{u}^{(s)}\right)}{\partial \boldsymbol{u}} = \left(-2 \sum_{i=1}^{I} \sum_{k=1}^{K} \mathcal{B}^\top(\bar{L}_k) \left(L_{i,k} - \mathcal{B}(\bar{L}_k)\boldsymbol{u}^{(s)}\right)\right)^\top,$$

and
$$B = \frac{\partial^2 Q\left(\boldsymbol{u}^{(s)}\right)}{\partial(\boldsymbol{u})^\top \partial \boldsymbol{u}} = 2\sum_{i=1}^{I}\sum_{k=1}^{K} \mathcal{B}^\top(\bar{L}_k)\mathcal{B}(\bar{L}_k).$$

An optimal solution for $\boldsymbol{u}$ can be computed by starting with $\boldsymbol{u}^{(0)} = \boldsymbol{0}$, iteratively solving (2.18) under the constraint (2.17) and updating $\boldsymbol{u}^{(s+1)} = \boldsymbol{u}^{(s)} + \boldsymbol{\delta}^{(s)}$ until $Q\left(\boldsymbol{u}^{(s)}\right)$ converges. An alternative approach to monotone smoothing with quadratic B-splines can be found in He & Shi (1998).

2.4 Approximate Dynamic Factor Models

In applications in multivariate statistics it is often desirable to reduce the dimension of the original dataset and to thereby compress the information and structure contained in the raw data. Factor models form a technique that has been widely employed for this purpose. By incorporating time series dynamics into factor models they became more and more interesting for economic applications, as well. In the firts part of this section an overview of the development of (approximate) dynamic factor models will be given. Then two different approaches to factor estimation within this framework will be discussed in Sections 2.4.2 and 2.4.3. In the last part of this section some light will be shed on the difference between principal components analysis and exploratory factor analysis as in practise factors are often estimated by principal components.

2.4.1 History and Recent Advances

The roots of dynamic factor models go back to the year 1904 when Charles Spearman found out that school children's results on a wide range of seemingly unrelated tasks were positively correlated (see Spearman, 1904, a brief review of the early history of classical factor analysis can be found e.g. in Steiger, 1979). He introduced a single common factor model and showed that it fitted his data well. He used this common factor which he called the g-factor as a measure for the unobservable mental ability or intelligence of a person. 15 years later Garnett (1919) extended Spearmans approach

2 Theoretical Background

to an M-factor model allowing for more than a single common component influencing a set of response variables. The "multiple factor analysis" gained popularity during the 1940's mainly due to Thurstone (1947) and became a commonly known and applied method in psychology and sociology. Let $\boldsymbol{x}_t = (x_{1,t}, \ldots, x_{N,t})^\top$ denote the realisations of N random variables $X_{i,t}, i = 1, \ldots, N, t = 1, \ldots, T$. In the classical M-factor model this multivariate time series is decomposed into M common and N idiosyncratic factors:

$$\boldsymbol{x}_t = \boldsymbol{\Lambda} \boldsymbol{f}_t + \boldsymbol{\epsilon}_t, \quad t = 1, \ldots, T,$$

where $\boldsymbol{\Lambda}$ is an $N \times M$-matrix of factor loadings, $\boldsymbol{f}_t = (f_{1,t}, \ldots, f_{M,t})^\top$ is the vector of the M common factors and $\boldsymbol{\epsilon}_t = (\epsilon_{1,t}, \ldots, \epsilon_{N,t})^\top$ is a vector of N idiosyncratic factors or disturbances. It is assumed that the $N + M$ common and idiosyncratic factors are mutually independent and uncorrelated across all t which seems to be a feasible assumption in most cross-sectional studies in social sciences. This assumption was the main obstacle which prevented the classical factor model from being applied to time series. Here, both types of factors will show some correlation over t.

In the economic context factor models were of great interest as they permitted to capture the main part of the variability of a large dimensional dataset with only a few common factors. This, in particular, was meaningful in forecasting models where the number of parameters to be estimated could be considerably reduced. This, in turn, would result in a significant reduction of uncertainty caused by the estimation of the unknown parameters. The stability of forecasting models with many predictors could be enhanced that way. Geweke (1977) and Sargent & Sims (1977) were the first who introduced time series dynamics into factor models. They also created the expression "dynamic factor analysis" and "dynamic factor model". Geweke loosened the restriction mentioned above by allowing for correlation of both, common and idiosyncratic factors over time t. He retained the restriction of $\boldsymbol{f}_t$ and $\boldsymbol{\epsilon}_t$ being mutually independent and postulated that both were covariance stationary and strictly indeterministic. By giving a frequency domain representation he showed that the dynamic factor model is especially suitable for time series which have most of their variation at low frequencies. He pointed out that this is the case for most macroeconomic series where features of interest like

2 Theoretical Background

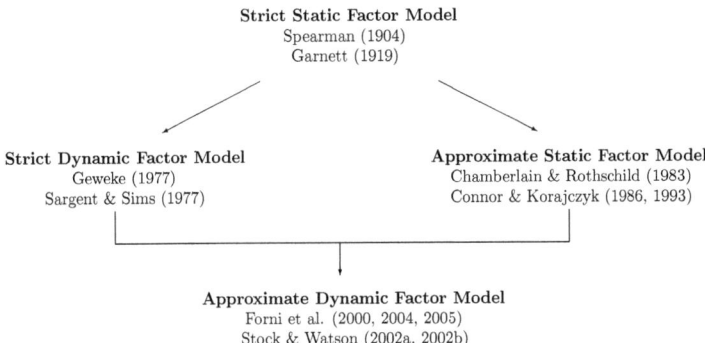

Figure 2.2: Schematic of the development of the approximate dynamic factor model.

business cycles have a six month or even longer frequency. Because of this dynamic factor models became interesting for scientists of the economic field.

Chamberlain (1983) and Chamberlain & Rothschild (1983) extended the classical "strict" factor model in another direction. They pointed out that the assumption of uncorrelated idiosyncratic factors is very unlikely in most applications in economics and finance. They therefore introduced the "approximate factor model" where the idiosyncratic distrubances are allowed to be weakly cross-sectionally correlated, i. e. the assumption

$$E(\epsilon_t^i \cdot \epsilon_s^j) = 0, \forall\ i,j,t,s \text{ with } i \neq j,$$

is being relaxed. The idea of the approximate factor structure was also taken up by Connor & Korajczyk (1986, 1993).

The first who formulated a combination of both mentioned extensions of the classic factor model were Forni & Reichlin (1998) and Forni et al. (2000). Their "generalized

2 Theoretical Background

dynamic factor model" is the synthesis of the concepts suggested by Geweke (1977) and Chamberlain & Rothschild (1983) and allows for serial correlation of both, common and idiosyncratic factors as well as for weak cross-correlation among the idiosyncratic part. However, they proposed to estimate the latent factors via dynamic principal components (see Brillinger, 1981) which is a two-sided estimation routine that includes leads and lags. Therefore, their approach was not apt for forecasting. Stock & Watson (2002a, 2002b) were the first who came up with a one-sided estimation routine using only contemporaneously available values of the predictors of interest. Their work focused on forecasting macroeconomic variables when a large number of candidate predictor variables is available. The number of time series might even exceed the number of observations in the dataset. Instead of excluding less relevant variables from the analysis they employ an approximate dynamic factor model where they estimate the factors using principal component analysis. They impose a vector-autoregressive structure on the common part and, if needed, univariate autoregressive structures on the idiosyncratic factors. An alternative one-sided estimation routine for the factors is proposed by Forni, Hallin, Lippi & Reichlin (2005). Similar to their earlier publications (Forni, Hallin, Lippi & Reichlin, 2000, 2004) they pursue a spectral density based approach and estimate the factors with generalized principal components what yields asymptotically better results as the method suggested by Stock & Watson (2002a) which they proof with the help of a simulation study and a real data example.

In the very recent past Byeong et al. (2009) in line with a number of other studies (see references therein) presented a semiparametric approach to dynamic factor models which they call the "dynamic semiparametric factor model" (DSFM). The main difference to the former mentioned techniques is that they let the factor loadings be semiparametric functions of some observable influencing variables. They use a multivariate Newton-Raphson-Algorithm to estimate the factor scores. Unlike Stock & Watson and Forni et al. they do not make use of a normalization in order to identify the common factors uniquely (up to the sign) but they show with the help of a simulation study that for any set of estimated factors there exists a transformed solution which has the same covariance structure as the original set of common factors and hence inference can be done based on any feasible solution. Another application of the DSFM is given in Borak & Weron (2008).

Up to now the publications of Stock & Watson (see Section 2.4.2 for more details) and Forni et al. (see Sectoin 2.4.3) are considered to be the main references for applications of approximate dynamic factor models in the field of economics and finance. A graphical sketch of the most relevant development stages of this type of factor model is given in Figure 2.2.

2.4.2 Common Factor Estimation via Principal Components Analysis

Stock & Watson (2002a) point out that for many macroeconomic forecasting problems economists monitor a large number of candidate predictor variables. Often the number of predictors even exceeds the number of observations available which makes a direct regression of the variable of interest on the predictors impossible. Instead of excluding less relevant variables from the analysis they formulate the idea that the economy is driven by some unobservable forces and the hundreds or maybe thousands economic variables are measures which contain information about a mixture of these driving forces. They estimate the forces using an approximate dynamic factor model. The forecasting setting is then reduced to a feasible size by plugging in an appropriate number of common factors instead of a huge number of predictor variables. In order to produce a k-step ahead forecast of a response time series y_t they suggest a two-step estimation through a static factor model for the predictors $\boldsymbol{x}_t = (x_{1,t}, \ldots, x_{N,t})^\top$

$$\boldsymbol{x}_t = \boldsymbol{\Lambda} \boldsymbol{f}_t + \boldsymbol{\epsilon}_t, \tag{2.19}$$

where $\boldsymbol{f}_t = (f_{1,t}, \ldots, f_{M,t})$ is a vector of M common factors, $\boldsymbol{\Lambda}$ is the $N \times M$-matrix of factor loadings and $\boldsymbol{\epsilon}_t$ the N-vector of idiosyncratic factors, together with a forecasting equation

$$y_{t+k} = \boldsymbol{f}_t^\top \boldsymbol{\beta}_f + \boldsymbol{w}_t^\top \boldsymbol{\beta}_w + e_{t+k}. \tag{2.20}$$

Here, $\boldsymbol{w}_t$ is a vector of covariates that shall be incorporated directly into the forecasting process, $\boldsymbol{\beta}_f$ and $\boldsymbol{\beta}_w$ are coefficient vectors and e_{t+k} is the forecasting error. The dynamic

factor model
$$x_{i,t} = \boldsymbol{\lambda}_i(L)\boldsymbol{f}'_t + \epsilon_{i,t}, \qquad (2.21)$$
and
$$y_{t+k} = (\boldsymbol{f}'_t)^\top \boldsymbol{\beta}_f(L) + \boldsymbol{w}_t^\top \boldsymbol{\beta}_w + e_{t+k}, \qquad (2.22)$$
with L as lag operator, can be rewritten in the static form (2.19) and (2.20) by setting $\boldsymbol{f}_t = ((\boldsymbol{f}'_t)^\top, (\boldsymbol{f}'_{t-1})^\top, \ldots, (\boldsymbol{f}'_{t-q})^\top)^\top$ if the lag polynomials are of finite order q. This is of great importance because for the static form the factors can be consistently estimated via principal component analysis for $N, T \to \infty$ as shown in Stock & Watson (2002a). For the identification of the factors (up to the sign) Stock and Watson impose the following assumptions:

- $(\Lambda^\top \Lambda / N) \to \boldsymbol{I}_M$,

- $\mathrm{E}(\boldsymbol{f}_t \boldsymbol{f}_t^\top) = \Sigma_{ff}$ is a diagonal matrix with $\sigma_{ii} > \sigma_{jj}$ for $i < j$,

- $T^{-1} \sum_t \boldsymbol{f}_t \boldsymbol{f}_t^\top \xrightarrow{p} \Sigma_{ff}$,

- $|\lambda_{i,j}| \leq c < \infty$ for some constant c.

That is, they postulate that each of the common factors significantly contributes to the explanation of the total variance in $\boldsymbol{x}_t$ and that the factor process $\boldsymbol{f}_t$ is covariance stationary. Furthermore, they allow for correlation among the idiosyncratic part in the sense of Chamberlain & Rothschild (1983) and Connor & Korajczyk (1986, 1993) by assuming

- $\mathrm{E}(\boldsymbol{\epsilon}_t^\top \boldsymbol{\epsilon}_{t+s}/N) = \gamma_{N,t}(s)$ with $\lim_{N\to\infty} \sup_t \sum_{s=-\infty}^{\infty} |\gamma_{N,t}(s)| < \infty$,

- $\mathrm{E}(\epsilon_{i,t} \cdot \epsilon_{j,t}) = \tau_{ij,t}$ with $\lim_{N\to\infty} \sup_t N^{-1} \sum_{i=1}^N \sum_{j=1}^N |\tau_{ij,t}| < \infty$,

- $\lim_{N\to\infty} \sup_t \sum_{i=1}^N \sum_{j=1}^N |\mathrm{Cov}(\epsilon_{i,s} \cdot \epsilon_{i,t}, \epsilon_{j,s} \cdot \epsilon_{j,t})| < \infty$.

The first assumption implies serial correlation among the idiosyncratic factors, the second allows them to be weakly cross-correlated and the third limits the fourth moments.

2 Theoretical Background

Stock & Watson (2002a, 2002b) thereby suggest a combination of the concepts of Geweke (1977) and Chamberlain & Rothschild (1983). The latter proposed to estimate factors via principal components analysis for the static case. An alternative approach using generalized principal components will be presented in the next section.

2.4.3 Common Factor Estimation via Generalized Principal Components

In the approximate dynamic factor model suggested by Forni, Hallin, Lippi & Reichlin (2000) the common factors were estimated using dynamic principal components analysis (see Brillinger, 1981). This estimation routine is two-sided meaning that both, leads and lags, are included which is no problem "in the middle" of the sample but it is not applicable to the first and last observations. Therefore, this method is improper for forecasting where factors of the most recent observations have to be determined.

Forni, Hallin, Lippi & Reichlin (2005) suggest a one-sided estimation routine for an approximate dynamic factor model. Following Stock & Watson (2002a, 2002b) they point out that a dynamic factor model of the form (2.21) and (2.22) with a finite lag structure can be written in the static form (2.19) and (2.20) which allows to use principal components for factor estimation. However, they argue that instead of standard principal components there may be better linear combinations that result in a better approximation of the space of common factors. As an alternative they recommend to use generalized principal components. The static form is necessary for the factor estimation through principal components. By imposing similar assumptions as Stock & Watson (2002a) they derive their estimation routine from the spectral density $\Phi(\theta)$. Let $\hat{\Phi}(\theta)$ denote the lag-window estimator of $\Phi(\theta)$ (that can be a two-sided mid-sample estimator). Spectral density estimates of the common and idiosyncratic factors can then be obtained by setting

$$\hat{\tilde{\Phi}}_f(\theta) = \hat{\nu}_1(\theta)\hat{p}_1(\theta)\hat{p}_1^*(\theta) + \cdots + \hat{\nu}_M(\theta)\hat{p}_M(\theta)\hat{p}_M^*(\theta),$$

and

$$\hat{\Phi}_\epsilon(\theta) = \hat{\nu}_{M+1}(\theta)\hat{p}_{M+1}(\theta)\hat{p}_{M+1}^*(\theta) + \cdots + \hat{\nu}_N(\theta)\hat{p}_N(\theta)\hat{p}_N^*(\theta),$$

2 Theoretical Background

where $\hat{\nu}_i$ is the i-th largest estimated eigenvalue of $\hat{\boldsymbol{\Phi}}(\theta)$ and $\hat{\boldsymbol{p}}_i(\theta)$ the corresponding eigenvector and the superscript $*$ denotes the transposed, complex conjugated version. M denotes the number of eigenvalues that are needed to capture the desired part of the variation in $\hat{\boldsymbol{\Phi}}(\theta)$. Using these results the covariance matrices of common and idiosyncratic factors are given by

$$\tilde{\boldsymbol{\Sigma}}_f = \int_{-\pi}^{\pi} \hat{\boldsymbol{\Phi}}_f(\theta) \mathrm{d}\theta,$$

and

$$\tilde{\boldsymbol{\Sigma}}_\epsilon = \int_{-\pi}^{\pi} \hat{\boldsymbol{\Phi}}_\epsilon(\theta) \mathrm{d}\theta.$$

The $N \times M$ matrix of factor loadings is build from the first M generalized eigenvectors of the matrices $\tilde{\boldsymbol{\Sigma}}_f$ and $\tilde{\boldsymbol{\Sigma}}_\epsilon$, that is, the solutions of the generalized eigenvalue problem

$$\tilde{\boldsymbol{\Sigma}}_f \boldsymbol{\lambda}_i = \rho_i \tilde{\boldsymbol{\Sigma}}_\epsilon \boldsymbol{\lambda}_i, \quad i = 1, \ldots, N,$$

where ρ_i denotes the i-th largest generalized eigenvalue and $\boldsymbol{\lambda}_i$ the corresponding eigenvector of the matrix couple $\tilde{\boldsymbol{\Sigma}}_f$ and $\tilde{\boldsymbol{\Sigma}}_\epsilon$ under the normalization constraints

$$\boldsymbol{\lambda}_i^\top \tilde{\boldsymbol{\Sigma}}_\epsilon \boldsymbol{\lambda}_i = \begin{cases} 1, & i = j, \\ 0, & i \neq j. \end{cases}$$

The factor loading matrix is then given by $\boldsymbol{\Lambda} = (\boldsymbol{\lambda}_1, \ldots, \boldsymbol{\lambda}_M)$.

In the last part of this section the differences between exploratory factor analysis and principal components analysis will be pointed out and it will be shown under which circumstances they yield approximately the same results.

2.4.4 Principal Components Analysis vs. Exploratory Factor Analysis

The aim of this section is to emphasize that principal components analysis (PCA) and exploratory factor analysis (EFA) are, although related, different methods and in general their results are not the same. In practise, they are often confused or used equivalently. It shall be explained under which circumstances both techniques yield quite similar

2 Theoretical Background

results. Instructions how to employ both techniques and information about the assumptions being made can be found in most textbooks on multivariate analysis, for example, Mardia, Kent & Bibby (1979) or Rencher (2002).

Although PCA and EFA are closely related and in practise even often confused they are not identical. Both procedures can be used to reduce the dimension of a dataset. They differ by the amount of variance which is accounted for in their models. In principal components analysis all the variance which is contained in N observed variables is preserved by N factors. The PCA model is

$$x_{i,t} = \lambda_{i,1} f_{1,t} + \lambda_{i,2} f_{2,t} + \cdots + \lambda_{i,N} f_{N,t},$$

with $t = 1, \ldots, T$ and $i = 1, \ldots, N$, where $x_{i,t}$ is the mean-corrected value of the t-th observation on the i-th random variable, $\lambda_{i,m}$ is the weight of the i-th variable on the n-th factor $f_{n,i}$ ($n = 1, \ldots, N$). The factors $f_{m,i}$ are assumed to be uncorrelated.

The EFA model only accounts for the amount of variance which is shared by all observed variables. Here, the variable $x_{i,t}$ has to be not only mean-corrected but standardized. The factor model can be written as

$$x_{i,t} = \lambda'_{i,1} f'_{1,t} + \lambda'_{i,2} f'_{2,t} + \cdots + \lambda'_{i,M} f'_{M,t} + \epsilon_{i,t}, \tag{2.23}$$

with $t = 1, \ldots, T$ and $i = 1, \ldots, N$ and $M \leq N$. Here N random variables are explained by usually less the N common factors and $\epsilon_{i,t}$ are the idiosyncratic disturbances. Without loss of generality it can be assumed that both common and unique factors have zero mean and unit variance as they are unknown in practice. Moreover, the unique factors are assumed to be independent of each other and of the common factors.

To point out the differences of PCA compared to EFA assume that the components are ordered by their corresponding eigenvalues, i. e. by the amount of variance explained through each component in decreasing order. When the dimension of a dataset shall be reduced by applying a PCA this can be done by retaining only the leading M components $f_{1,t}, \ldots, f_{M,t}$ and dropping the information contained in the remaining $N - M$ components. This leads to

$$x_{i,t} = \sum_{m=1}^{M} \lambda_{i,m} f_{m,t} + e_{i,t},$$

2 Theoretical Background

which appears to be similar to (2.23). A closer look at the residuals $e_{i,t}$ reveals the difference:

$$e_{i,t} = \sum_{m=M+1}^{N} \lambda_{i,m} f_{m,t}.$$

In the factor model we presumed that the unique factors $\epsilon_{i,t}$ were mutually independent. This assumption is violated in the PCA model. Here $e_{i,t}$ and $e_{j,t}$ are not independent for $i \neq j$ as the same $f_{m,t}$'s are involved.

Chamberlain & Rothschild (1983) show for their approximate static factor model that PCA and EFA are asymptotically equivalent for $N, T \to \infty$. Stock & Watson (2002a) demonstrate the same for the approximate dynamic factor model. However, there are situations where both, PCA and EFA, yield approximately the same results for finite N and T, as well. This is the case when the communalities of the EFA model are close to unity. The communalities $h_i^2, i = 1, \ldots, N$ are defined as

$$h_i^2 = \sum_{m=1}^{M} (\lambda'_{i,m})^2.$$

If they are close to unity this means that the main part of the variance in the N observed variables is due to the M common factors and that the unique factors are of little importance. In this case in a PCA the first M components will explain the main part of the variance, as well, and the results of both methods will only differ slightly.

In this chapter the main important theoretical concepts have been presented that will be applied in the following part of this thesis. The focus thereby was on landmark-based curve registration and approximate dynamic factor models because these methods will be employed in new frameworks. In Chapter 3 the former will be used to estimate time-warping functions for parts of an online monitored time series. The latter will find application in high-resolution forecasting of water temperature (Chapter 4) and energy demand (Chapter 5).

3 Application: Landmark Specification in Water Temperature Data

If we compare meteorological air temperature measurements taken over different years in the European climate zone we usually find the warm and the cold period of a year, i.e. summer and winter, clearly exposed. These yearly temperature curves not only differ in amplitude meaning that we find hotter summers in some years and colder ones in others. The periods may also be shifted in time, i.e. we can observe that in some years the warm or cold period of the year starts earlier, lasts longer or the contrary, respectively. These time shifts can be called phase variation.

Due to the physical heat transfer the patterns carry over to river water temperature measurements. As pointed out in detail in Section 1.1 such shifts have an influence on the fish population in terms of, for instance, migration, spawning or maturing of juvenile fish, see e.g. Ovidio et al. (2002) or Rakowitz et al. (2008) and references therein. As mentioned in Section 1.1 there are certain water temperature thresholds during the different stages of the spawning cycle of fish that must not be crossed. Furthermore, water temperature among other things is a significant stimulus and trigger for spawning and migration, see e.g. Ovidio et al. (1998). Such triggers could be considered as landmarks in the annual variation of water temperature. Given a set of landmarks that can be reliably found in every year we can judge whether in a particular year a season is running behind or ahead of the "average" time scale. In this application, however, we focus on the problem of finding landmarks from a statistical perspective by looking for recurrent events in temperature data. We will also investigate if the conclusions drawn from these landmarks can be linked to the ecological versions consisting of triggers and

3 Application: Landmark Specification in Water Temperature Data

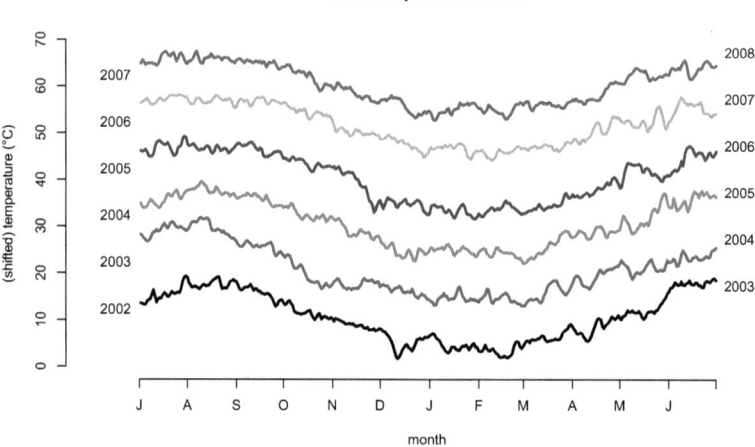

Figure 3.1: Daily average water temperature recorded from 1 July 2002 to 30 June 2008. For a better visualization the curves from the second year on have been shifted by cumulatively adding 10°C.

spawning stages.

Note that the "average year" which we want to use for our analysis has to be defined from the landmarks found in the dataset. For any landmark we therefore take the rounded mean of its appearance in the differnt years of our data sample as reference points. These form the reference year that will be used to evaluate deviations for a particular year in terms of stretching or compressing time.

The data at hand consist of hourly water temperature measurements from the river Wupper in the North-Western part of Germany. The measurements were taken upstream of the city of Wuppertal where two fossil-fueled powerplants use the water as cooling device (see also Section 1.1). Corresponding hourly air temperature readings from the

3 Application: Landmark Specification in Water Temperature Data

area of interest are available, as well. The water temperature data can be considered as free from any kind of (relevant) human interference. The curves of the maximum daily water temperatures are shown in Figure 3.1 with observations being shifted vertically (by cumulatively adding 10°C) for better visual impression. Years are fixed to last from 1 July to 30 June instead of the Julian calendar, which is more coherent with the spawning cycle of Salmonidae that starts in autumn and ends in late winter / early summer. Just by pure visual inspection we see a number of features. For instance, summer 2003 lasted long (until September) and was followed by an early spring in 2004. Our intention is to get statements like these more formally based on statistical grounds.

In order to obtain an indication whether a season is running ahead or behind the average year we want to employ a technique known as "curve registration" from the field of functional data analysis. We thereby closely follow the suggestions of Kneip & Gasser (1992) and Gasser & Kneip (1995) who present a set of groth acceleration curves of children that show the same features, i.e. the same extreme points but shifted in time. These features are called "structural functionals" or "landmarks". With knowledge about the locations of these characteristics a so-called "time-warping function" can be derived for every functional observation that aligns the acceleration curves to a priori defined reference points by compressing or expanding time. A detailed review of landmark-based time-warping as well as an overview of related techniques can be found in Section 2.3. However, the data structure and the focus of our data analysis does not allow for a direct application of these methods because of the following reasons. First, looking at our data it appears that the only clearly exposed local extrema mark summer and winter and other local extrema appear to be more or less random events. Hence, the underlying ideas of finding local extrema as suggested in Gasser & Kneip (1995) seems not fruitful except of defining these two yearly extrema. Secondly, and more importantly, all methods described in Section 2.3 are designed for a retrospective point of view, that is, the complete data are necessary to run a warping or registration procedure. In our example this would mean, based on the data of a year we could retrospectively decide whether seasons were running ahead or behind the "average year". We want however an procedure reacting "online" by looking at recent measurements.

3 Application: Landmark Specification in Water Temperature Data

Hence, based on data collected in the progression over a year we want to decide whether a landmark has been reached. Our intention is to find landmarks in a data driven style and to retrieve structures which can be found quite reliably every year. To do so, we will make use of running means, temperature thresholds, principal component analysis and canonical correlation relating water temperature to air temperature. This shows similarities to Silverman (1995) but instead of looking at the average water temperature shown in Figure 3.1 we also look at the daily variation of the temperature. We will see that the data are quite informative and provide relevant information about the course of the seasons.

In the Section 3.1 we will present four online methods for the specification of landmarks. These will be used in Section 3.2 for time-warping. Here we will also explain in detail which modifications to the classical techniques presented in Section 2.3 are necessary to handle the time series structure of our data. The results will be linked to data concerning the fish reproduction cycle of the Brown Trout in the river Wupper. In Section 3.3 we investigate the variability of the different landmarks by considering their bootstrapped distribution before we summarize our results in Section 3.4.

3.1 Landmark Specification

We will subsequently introduce landmarks which are found in different ways by analyzing water temperature and its connection to air temperature. However, before going into detail we first introduce our notation. Let $\boldsymbol{w}_t = (w_{t,0}, \ldots, w_{t,23})$ be the vector of hourly water temperature at time t, where t can be expressed by $t = (i, d)$ with i indicating the year and d giving the day of the year (yearday). Analogously, we define the vector of hourly air temperatures $\boldsymbol{a}_t$. We decompose both temperature vectors into

$$\boldsymbol{w}_t = \boldsymbol{1} \cdot \bar{w}_t + \boldsymbol{x}_t \qquad (3.1)$$

$$\boldsymbol{a}_t = \boldsymbol{1} \cdot \bar{a}_t + \boldsymbol{b}_t \qquad (3.2)$$

where $\boldsymbol{1}$ is a vector of 1's, $\bar{w}_t$ and $\bar{a}_t$ are the average water and air temperature at day d in year i and $\boldsymbol{x}_t$ and $\boldsymbol{b}_t$ are the remaining water and air temperature courses over that

3 Application: Landmark Specification in Water Temperature Data

day, respectively. Note that $\boldsymbol{x}_t = (x_{t,0}, \ldots, x_{t,23})$ has 24 elements summing up to zero, the same is true for $\boldsymbol{b}_t$.

3.1.1 Running Means

As natural set of landmarks in the data we have yearly maximum and minimum temperature, see Figure 3.1. These are, however, generally weakly exposed so that we do not make use of finding global extrema per year. Instead, we employ the simple strategy by using two running means and define their cut points as landmarks. To be specific, a first pair of landmarks is defined by the cut points of simple running 100 days / 200 days means.

The temperature curves shown in Figure 3.1 do not exhibit any further exposed functional characteristics. We therefore look at temperature thresholds.

3.1.2 Temperature Thresholds

In order ot fix these landmarks we apply simple t-tests to the daily mean temperature $\bar{w}_{\tilde{t}}$ in a small time window $\tilde{t} \in \{t - 14, \ldots, t\}$ and test the hypothesis

$$H_0 : E(\bar{w}_{\tilde{t}}) \geq \eta \quad \text{for} \quad \tilde{t} \in \{t - 14, \ldots, t\}, \tag{3.3}$$

against its one-sided alternative. As temperature thresholds we choose $\eta \in \{7°C, 11°C, 15°C\}$. We set the landmarks for temperature η every time the corresponding p-value of the test crosses the .1 threshold. The p-values for the three different values of η and the related landmarks are shown in Figure 3.3. In order to stabilize the location of the landmarks an additional restriction is needed here. We postulate that the minimum distance between two adjacent landmark for the same value of η is 30 days. In total this criterion results in three additional pairs of landmarks.

The landmarks so far are based on the average daily temperature. We will now look deeper into the daily variation of the temperature.

3 Application: Landmark Specification in Water Temperature Data

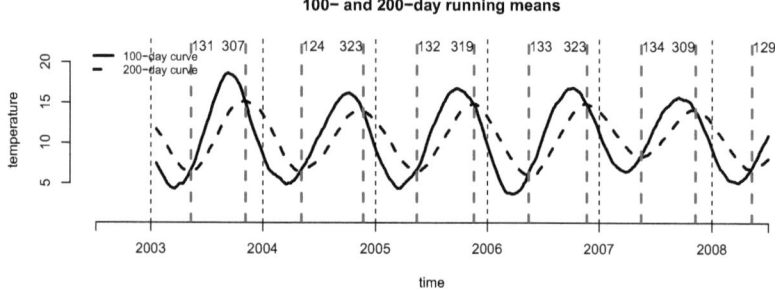

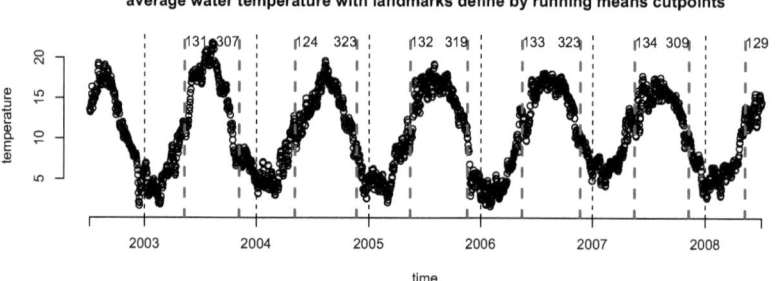

Figure 3.2: Top panel: 100- and 200-days running mean curves of the average temperature. The cut points which are selected as landmarks are marked by red vertical dashed lines. Bottom panel: Daily average water temperature. The red vertical dashed lines mark the position of the running mean based landmarks.

3 Application: Landmark Specification in Water Temperature Data

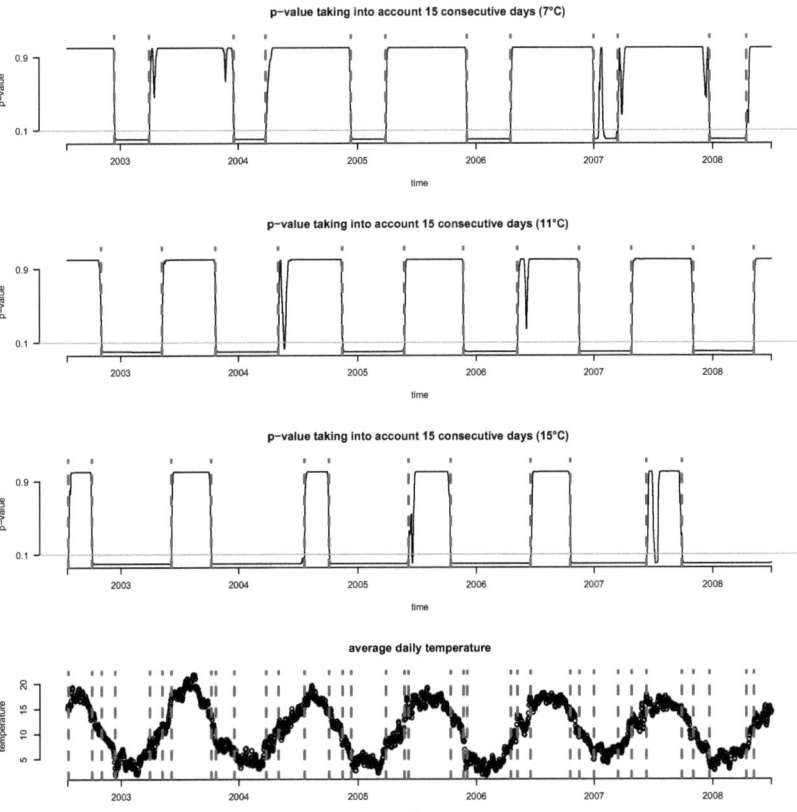

Figure 3.3: First three panels: p-values of the one-sided t-tests to hypothesis (3.3) for $\eta = 7°C, 11°C, 15°C$ taking into account 15 consecutive days. Bottom panel: Daily average water temperature. The vertical dashed lines mark the positions of the daily mean temperature landmarks for $\eta \in \{7°C, 11°C, 15°C\}$.

3 Application: Landmark Specification in Water Temperature Data

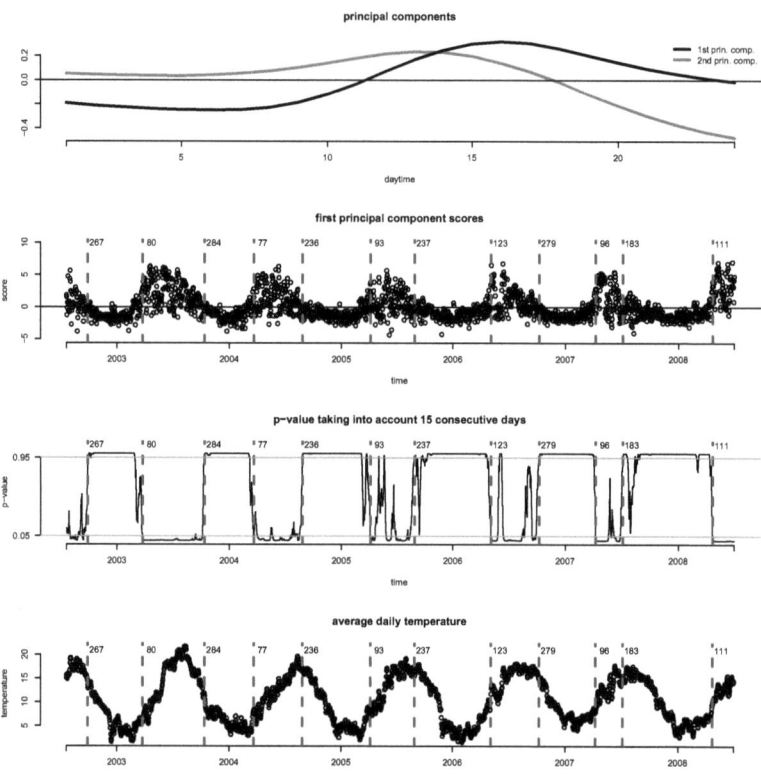

Figure 3.4: Top panel: The first two (smooth) principal components $\lambda_{x,k}(h)$, $k=1,2$ of the daily water temperature course. Second panel: Scores of the first principal component. Third panel: p-value levels resulting from (3.6). Bottom panel: Daily average water temperature. The vertical dashed lines mark the positions of the PCA-based landmarks.

3 Application: Landmark Specification in Water Temperature Data

3.1.3 Daily Temperature Curve

We now consider the daily temperature course $\boldsymbol{x}_t = (x_{t,0}, \ldots, x_{t,23})$ defined in (3.1). Note that $\boldsymbol{x}_t$ sums up to zero and does not provide any information about the particular level of the temperature but instead it gives its daily fluctuation. We first extract the hourly mean curve $\mu_x(h)$ by setting

$$x_{t,h} = \mu_x(h) + \bar{x}_{t,h}. \tag{3.4}$$

Here $\mu_x(h)$ is a smooth function in hour h which in fact is calculated taking the mean value of available observations, that is

$$\hat{\mu}_x(h) = \frac{1}{T} \sum_{t=1}^{T} x_{t,h},$$

with T as number of available observations. We then run a principal component analysis (PCA) to obtain

$$\bar{x}_{t,h} = \sum_k \lambda_k(h) y_{t,k} \tag{3.5}$$

where $\lambda_k(\cdot)$ is the k-th (smooth) principal component and $y_{t,k}$, $k = 1, 2, \ldots$ are uncorrelated scores with mean 0 and variance σ_k^2. Decomposition (3.5) could be carried out with functional principal component analysis (Ramsay & Silverman, 2005) as implemented in the fda package in R. Based on the amount of data available, however, a standard PCA applied to $\bar{x}_{t,h}$ also provides smooth, functional principal components. We may consider the principal components as normed orthogonal functions, i.e. $\int \lambda_k(h) \cdot \lambda_l(h) \, dh = \delta_{kl}$ with $\delta_{kl} = 1$ for $k = l$ and $\delta_{kl} = 0$ otherwise. The assumption $E(y_{t,k}) = 0$ holds by construction over the entire data, that is when averaging over the entire dataset. Within a year, however, some annual fluctuations of $E(y_{t,k})$ around 0 become visible. In the top panel of Figure 3.4 we show the first two fitted principal components $\hat{\lambda}_k(h)$, $k = 1, 2$ corresponding to the largest two eigenvalues which cover 91% of the variability (first component: 83%). The first principal component carries the daily temperature variation with a minimum temperature in the early morning and a maximum in the late afternoon. The corresponding fitted score $\hat{y}_{t,1}$ for the first component is shown in the

3 Application: Landmark Specification in Water Temperature Data

second panel of Figure 3.4. Obviously, it carries an annual structure showing periods of the year with strictly negative values and low variation (winter) while in other periods the score varies around zero. This structure will now be exploited to define a further pair of landmarks.

To determine the landmarks we consider the scores of the first component and primarily check the sign of the scores and whether it changes form "−" (winter) to "+" (summer). Hence, the intention is to locate the structural break in the first score. This is done by making use of a local one-sided t-test taking into account the foregoing 15 days. That is, we test the null-hypothesis

$$H_0 : E(y_{\tilde{t},1}) \leq 0 \quad \text{for} \quad \tilde{t} \in \{t-14,\ldots,t\}, \tag{3.6}$$

against the one-sided alternative. Figure 3.4 shows the resulting p-values which give a clear separation between the seasons by either taking large or small values. We locate the first landmark when the p-value drops below .05 for the first time in the year. A second landmark is set when the p-value rises again over .95. The p-value levels in Figure 3.4 clearly show the annual features although there are days within the high- or low-level periods, where the mentioned thresholds are crossed and no landmark should be declared. To stabilize the procedure we add the further condition and postulate that the minimum distance between two consecutive landmarks is 120 days. In the bottom panel in Figure 3.4 we show the location of the landmarks on the course of the (daily) average water temperature over the years. Note that even though the structure in the first principal score component is well exhibited, the landmarks do not specify any exposed structure in the yearly temperature curve itself. Nonetheless, it shows the days when the daily temperature variation changes.

A fourth landmark criterion will now be defined by looking at the correlation between the daily temperature courses of water and air temperature.

3.1.4 Correlation between Water and Air Temperature

As final set of landmarks we look at the relation between the daily courses of air and water temperature. We therefore seek for a pair of landmarks which occur in the correlation

3 Application: Landmark Specification in Water Temperature Data

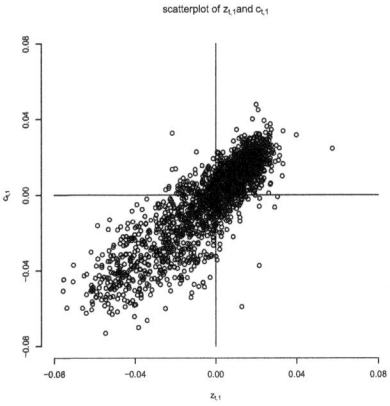

Figure 3.5: Scatterplot of $z_{t,1}$ plotted against $c_{t,1}$.

between these two measurements. In particular, we make use of (functional) canonical correlation and consider the mean corrected water and air temperature $\bar{x}_{t,h}$ and $\bar{b}_{t,h}$, respectively, with $b_{t,h} = \mu_b(h) + \bar{b}_{t,h}$ where $\mu_b(h)$ is the hourly mean structure extracted from $b_{t,h}$ in the same way as we obtained $\mu_x(h)$ in (3.4). We decompose

$$\bar{x}_{t,h} = \sum_k \zeta_k(h) \, z_{t,k} + \epsilon_{(x),t,h}, \quad \bar{b}_{t,h} = \sum_k \gamma_k(h) \, c_{t,k} + \epsilon_{(b),t,h}, \qquad (3.7)$$

where $\zeta_k(h)$ and $\gamma_k(h)$ are the canonical correlation functions fulfilling the orthonormality constraints $\int \zeta_k(h) \cdot \zeta_l(h) \, dh = \int \gamma_k(h) \cdot \gamma_l(h) \, dh = \delta_{kl}$ with $\delta_{kl} = 1$ for $k = l$ and $\delta_{kl} = 0$ otherwise. The score vectors $z_t = (z_{t,1}, z_{t,2}, \dots)$ and $c_t = (c_{t,1}, c_{t,2}, \dots)$ can be considered as random with $\text{Var}(z_t) = I$, $\text{Var}(c_t) = I$ and $\text{Cov}(z_t, c_t) = \text{diag}(\rho_1, \rho_2, \dots)$, where the canonical correlations are ordered such that $\rho_1 > \rho_2 > \dots$. The correlation model (3.7) can be fitted using the implemented version in the fda package in R, see Ramsay & Silverman (2002) or Leurgans, Moyeed & Silverman (1993). The amount of data available, however, also allows to use a standard canonical correlation applied to

3 Application: Landmark Specification in Water Temperature Data

$\bar{x}_{t,h}$ and $\bar{b}_{t,h}$. Figure 3.6 (top panel) shows the first canonical functional components $\zeta_1(h)$ and $\gamma_1(h)$ for water temperature and air temperature, respectively. The course of the first canonical coefficient mirrors that the minimum air temperature is reached by about 6 a.m., while the minimum water temperature is reached by 8 a.m. Moreover, the afternoon air temperature is about equally useful to express the correlation with the maximum water temperature at around 4 p.m.

For the definition of the landmarks we look at the first canonical correlation and the fitted coefficients $z_{t,1}$ and $c_{t,1}$ which carry the maximum canonical correlation of order 0.87. Figure 3.5 shows a scatterplot of $z_{t,1}$ plotted against $c_{t,1}$ exhibiting the correlation structure. We find that points in the lower left quadrant belong to days during late spring and summer while the remaining three quadrants contain observations from all over the year. This structure can be better exploited by defining the modified score that only gives weight to data in the lower left sector:

$$Z_t = \begin{cases} z_{t,1} \cdot c_{t,1}, & \text{if } z_{t,1}, c_{t,1} < 0, \\ 0, & \text{otherwise.} \end{cases}$$

The daily scores Z_t are shown in the second panel of Figure 3.6 and the seasonal pattern becomes more obvious. To formalize the definition of the last pair of landmarks we, again, pursue a one-sided t-test now taking into account the 25 previous days. We test the hypothesis

$$H_0 : E(Z_{\tilde{t}}) \geq E(Z) \quad \text{for} \quad \tilde{t} \in \{t - 24, \ldots, t\},$$

against the one-sided alternative where $E(Z)$ is the mean value of Z_t averaged over t. For our example we replace $E(Z)$ by its empirical version which equals .58. A landmark is defined when the corresponding p-value (see third panel of Figure 3.6) crosses the threshold of 0.1 while a minimum distance of 50 days is postulated between two adjacent landmarks. The bottom panel illustrates the location of the landmarks in the water temperature course.

3 Application: Landmark Specification in Water Temperature Data

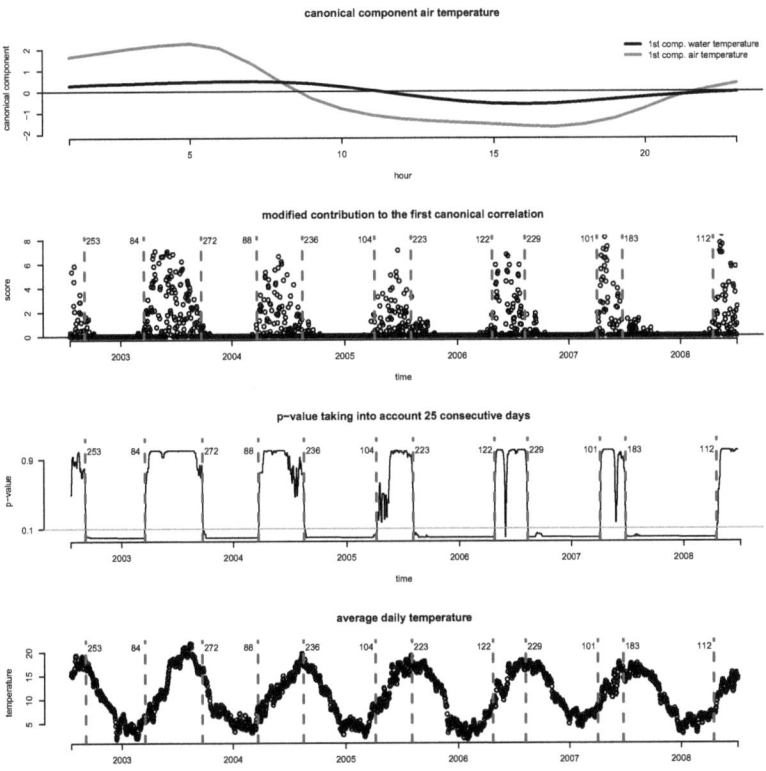

Figure 3.6: The first panel illustrates $\zeta_1(h)$ and $\gamma_1(h)$. Second panel: Modified contribution to the first canonical correlation: Z_t. Third panel: Level of p-values of a one-sided t-test accounting for 25 days. Bottom panel: Daily average water temperature. The vertical dashed lines mark the positions of the canonical correlation landmarks.

3 Application: Landmark Specification in Water Temperature Data

In this section we introduced four criteria build for specifying landmarks in a time series with regularly arriving new observations. We chose the cut points of running means as an equivalent to seasonal temperature extrema. The three remaining criteria are based on a one-sided t-test that only takes into account recent observation. The obtained landmarks will now be used to register the data in order to find time-warping functions that can be used as indicator whether a season is running early or late compared to a reference year.

3.2 Registering the Data

A major intention of our data exploration is to apply landmark-based curve registration as developed by Kneip & Gasser (1992) and Gasser & Kneip (1995). However, in our application we do not face independent functional observations but one single time series whose yearly courses shall be aligned to that of an average year. The registration method we pursue and the modifications to the original technique will be described in the first part of this section before we present our time-warping results. In the second part those results shall be linked to spawn cycle data of the Brown Trout (Salmo Trutta).

3.2.1 Landmark-based Time-Warping

We rewrite the temperature $w_{t,h}$ as $w_{i,h}(d)$ indicating the temperature (at hour h) in year i at day d. We assume that the course carries annual characteristics but may be shifted in time. We therefore need a (strictly) monotonically increasing time transformation $d \mapsto \pi_i(d)$ such that

$$w_h^*(d) := w_{i,h}(\pi_i(d)), \quad h = 0, \ldots, 23, \quad i = 1, 2, \ldots,$$

where $w_h^*(d)$ are so-called registered curves (see Ramsay & Silverman, 2005, page 132). The time transformation $\pi_i(d)$ works by accelerating or slowing down time, respectively, and is called the time-warping function, subsequently. To estimate functions $\pi_i(d)$ we

3 Application: Landmark Specification in Water Temperature Data

pursue the landmark-based approach suggested by Kneip & Gasser (1992) (see Section 2.3.2) which has to be extended to our application in so far that the functional observations $w_{i,h}(d)$ are not independent observations in i but parts of the same time series $\boldsymbol{w}_t$. Therefore, we derive a single time-warping function for the entire data

$$\tilde{\pi} : [0, T] \rightarrow [0, T],$$
$$t \mapsto \tilde{\pi}(t),$$

and define $\pi_i(d) := \tilde{\pi}(t)$, bearing in mind that time t is indicated by year i and day d. Furthermore, we do not require the curves $w_{i,h}(d)$ to be at least twice differentiable, although it would be possible by smoothing our data. In our context the landmark criteria described in Section 3.1 are functionals on $D([0,T])$, the set of all discrete functions over the time interval $[0,T]$. So that a mathematical definition of some landmark functional L could be given by

$$L : D([0,T]) \rightarrow [0,T] \cup \{\text{NA}\},$$
$$t \mapsto L(t) = L_i(d),$$

where the value NA ist taken if the landmark is missing in an observation.

Let $\{L_{i,k} | i = 1, \ldots, I, k = 1, \ldots, K\}$ denote the set of landmarks with index i giving the year and k indicating the landmarks resulting form the different criteria. For our example we have $K = 12$ possible landmarks per year. Note that $L_{i,k}$ may exceed the interval $[1, 365]$ which happens if a landmark located near the beginning or end of a year is shifted into the adjacent year. For each k we define the reference point

$$\bar{L}_k := \frac{1}{I} \sum_{i=1}^{I} L_{i,k},$$

giving the average landmark location over the years. We now consider the landmarks and their reference points in the context of the entire data series and set

$$\bar{\bar{L}}_{i,k} := (i-1) \cdot 365 \cdot \bar{L}_k \quad \text{and} \quad \tilde{L}_{i,k} := (i-1) \cdot 365 \cdot L_{i,k}.$$

This gives a set of $I \cdot K$ data points $(\bar{\bar{L}}_{i,k}, \tilde{L}_{i,k})$ with $i = 1, \ldots, I$ and $k = 1, \ldots, K$. The time-warping functions shall be constructed by applying monotone smoothing (see

3 Application: Landmark Specification in Water Temperature Data

Section 2.3.3) to the set of two-dimensional data points $(\bar{L}_{i,k}, \tilde{L}_{i,k})$, that is, instead of matching the observed landmarks exactly with their reference points we assume

$$\tilde{L}_{i,k} = \tilde{\pi}(\bar{L}_{i,k}) + \epsilon_{i,k},$$

where $\tilde{\pi}$ is a monotonically increasing time-warping function and $\epsilon_{i,k}$ is an error term. Introducing an error term is also advantageous as it allows us to ignore the consistency assumption among the landmarks which was postulated by Kneip & Gasser (1992). In practise it would be inprobable to only observe consistent landmarks when criteria are employed that focus on totally different characteristics. For practicability reasons we relaxed the strict monotonicity assumption on the time-warping functions and only consider them to be non-decreasing. We thereby follow the ideas of Kelly & Rice (1990). The exact smoothing procedure is presented in Section 2.3.3. Due to modifications of the warping procedure introduced above there are some adjustments to the smoothing equations necessary, as well. The sum of squares criterion (2.16) that has to be minimized changes to

$$Q(\boldsymbol{u}) := \sum_{i=1}^{I}\sum_{k=1}^{K}\left(\tilde{L}_{i,k} - \mathcal{B}(\bar{L}_{i,k})\boldsymbol{u}\right)^2,$$

and the matrices involved in the resulting quadratic form (2.18) have to be replaced by

$$\boldsymbol{a}^{(s)} := \frac{\partial Q\left(\boldsymbol{u}^{(s)}\right)}{\partial \boldsymbol{u}} = -2\sum_{i=1}^{I}\sum_{k=1}^{K}\mathcal{B}^\top(\bar{L}_{i,k})\left(\tilde{L}_{i,k} - \mathcal{B}(\bar{L}_{i,k})\boldsymbol{u}^{(s)}\right),$$

and

$$\boldsymbol{B} := \frac{\partial^2 Q\left(\boldsymbol{u}^{(s)}\right)}{\partial (\boldsymbol{u})^\top \partial \boldsymbol{u}} = 2\sum_{i=1}^{I}\sum_{k=1}^{K}\mathcal{B}^\top(\bar{L}_{i,k})\mathcal{B}(\bar{L}_{i,k}).$$

If a landmark is missing, i. e. it takes the value NA, the corresponding summand simply has to be dropped from the calculations.

The results are shown in Figure 3.7. If a season is running ahead of its correspondent in the average year we find $\pi_i(d)$ below the diagonal. A course above it means that the season is running late. For example the fall in 2003 came late and did not last long as it was followed by an early winter which in turn merged with an equally early spring and summer.

3 Application: Landmark Specification in Water Temperature Data

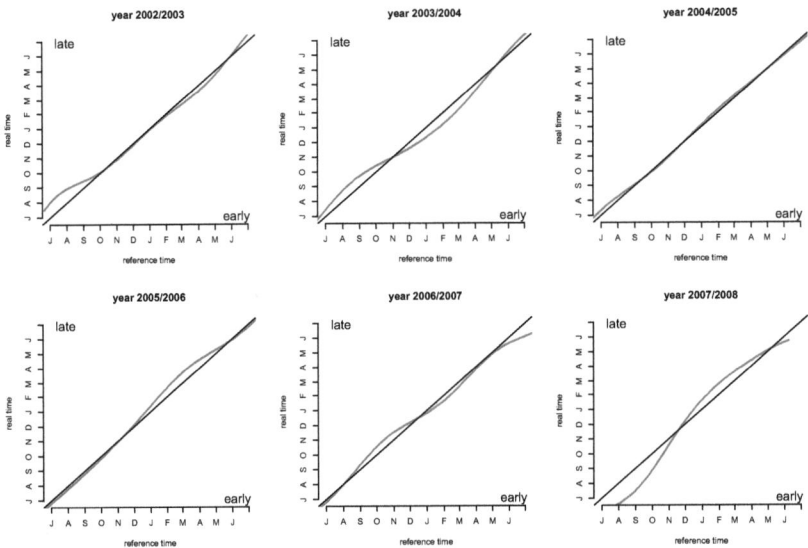

Figure 3.7: Course of the time-warping functions $\pi_i(d)$ for the six years of our dataset.

3.2.2 Linking to Ecological Data

We may try to connect our results with available ecological data in fish reproduction. Temperature pattern in rivers are considered as trigger for fish spawning, which can be explored by connecting the time warping function $\pi_i(d)$ to annual data on fish populations. We therefore use data containing the stages of the spawning cycle of Salmo Trutta (Brown Trout) for the last three years of data (July 2005 - June 2008) considered in the preceding section. The data trace from local records of fish surveillance in the upper part of the river Wupper. These stages can be taken as landmarks itself and we may consider the matching to the temperature based landmarks. In chronological order the fish go through the stages *Begin of Spawning Time*, *End of Spawning Time*, *Eye-Point Stage*,

3 Application: Landmark Specification in Water Temperature Data

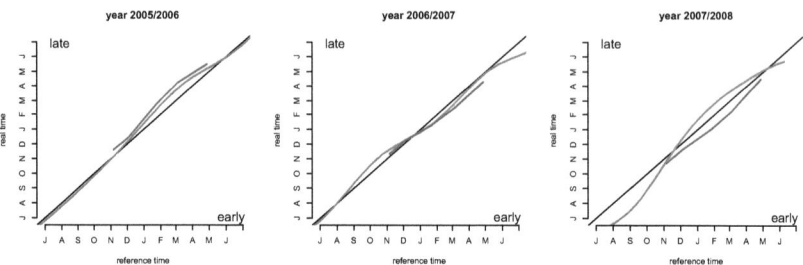

Figure 3.8: Spawn cycle phases of Salmo Trutta are indicated by red time-warping functions. Our results are given as green lines.

Eclosion and *Emergence*. Let $L_{i,k}^*$ be the beginning of the k-th reproduction stage in year i. We register the stages through linear interpolation of the points $(\bar{L}_k^*, L_{i,k}^*)$ where $\bar{L}_k^* := \frac{1}{I}\sum_{i=1}^{I} L_{i,k}^*$. The corresponding plots for the three considered years are shown in Figure 3.8. The time-warping functions $\pi_i(d)$ calculated above are plotted as green lines, the equivalents calculated from the spawn cycle dataset are given in red.

For the first two years (2005/2006 and 2006/2007) our landmarks in general coincide with the ecological candidates and the shape of the ecological time-warping functions matches those which we have calculated in a data driven way. Shift differences on the reference time axis may be caused by the small amount of ecological data that was available to calculate the reference points $\bar{L}_k^*$. For the third year (2007/2008) the matching does not work properly. We find that the spawn cycle takes place rather early in the year. This seems to be highly correlated with the very early fall found by our procedure. This could be taken as an indication that the spawning cycle of the Brown Trout is not only dependent on the current water temperature but on the temperature level in the foregoing months, as well.

3 Application: Landmark Specification in Water Temperature Data

In this section we used the landmarks defined in Section 3.1 to align the curves to their reference time points. We demonstrated what modifications were necessary to apply the classical landmark-based time-warping technique to our time series data and we compared our results to landmarks based on ecological data of the spawn cycle of the Brown Trout. We will now give insight into the variability of landmarks that is caused by the use of such different definition criteria and that may cause inconsistencies among the landmarks within a year.

3.3 Variability of Landmarks

The calculated landmarks itself are random variables and their specification is therefore stochastic. The amount of variability differs thereby between the landmarks. This is not surprising as, for instance, the running mean landmarks are based on a retrospective view on the 200 foregoing days which leads to a quite stable or robust behavior while the other landmarks are determined by accounting only for fifteen foregoing days. In order to assess the variability of landmark specification we therefore run a bootstrap procedure.

3.3.1 The Bootstrap Procedure

Bear in mind that all landmarks are calculated from derived time series data which are subsequently generally notated as e_t. For instance, the temperature threshold landmark (Section 3.1.2) calculates the landmark from $e_t = \bar{w}_t$ or the daily temperature curve landmark (Section 3.1.3) is calculated from $e_t = y_{t,1}$. Note that e_t can also be multivariate, like for the canonical correlation landmark (Section 3.1.4) with $e_t = (z_{t,1}, c_{t,1})$. The idea is now to bootstrap the series e_t to obtain information about the variability of the landmark specification. Note that e_t is not stationary, so that we do not resample the entire series but apply the bootstrap to a window $e_{t_0-\Delta}, \ldots, e_{t_0}$ where t_0 is the time point of interest where the landmark criterion shall be evaluated and $\Delta \leq 50$ is appropriately

3 Application: Landmark Specification in Water Temperature Data

chosen.

Bootstrapping in time series has been generally treated in Härdle, Horowitz & Kreiss (2003), see also Paparoditis & Politis (2002). We follow these ideas but first extract the non-stationarity by setting

$$e_t = \mu_e(t) + \tilde{e}_t,$$

where μ_e is mean trend over the interval $[t_0 - 50, t_0]$ and $\tilde{e}_t$ are the autocorrelated residuals. As we look at a rather small interval only, we assume μ_e to be locally linear and fit it with ordinary least squares. For the remaining residuals we fit an AR(p) process

$$\tilde{e}_t = \sum_{k=1}^{p} \boldsymbol{R}_k \tilde{e}_{t-k} + \epsilon_t,$$

where matrix $\boldsymbol{R}_k$ is the (matrix valued) autoregressive correlation. Residuals ϵ_t are possibly heteroscedastic which motivates to pursue a wild bootstrap see e. g. Mammen (1993). Therefore we draw ϵ_t^* from a two-point distribution given by $\hat{\epsilon}_t \cdot ((1-\sqrt{5})/2, (1+\sqrt{5})/2)$ with probabilities $((5+\sqrt{5})/10, (5-\sqrt{5})/10)$ where $\hat{\epsilon}_t$ is the empirical version of ϵ_t. Inserting the resulting bootstrap in the autoregressive models gives

$$\tilde{e}_t^* = \sum_{k=1}^{p} \hat{\boldsymbol{R}}_k \tilde{e}_{t-k}^* + \epsilon_t^*,$$

where $\hat{\boldsymbol{R}}_k$ is the fitted autocorrelation. The bootstrap sample of the original time series can finally be constructed by $e_t^* = \hat{\mu}_e(t) + \tilde{e}_t^*$. Based on such bootstrap replicates we can now calculate the bootstrap version of the landmarks.

The results of the application of this procedure will be demonstrated in the next part of this section.

3.3.2 Bootstrap Application

For all four landmark criteria we bootstrapped the underlying time series, that is, we have set $e_t = \bar{w}_t$ for the running means and temperature threshold criterion and $e_t = y_{t,1}$ for the PCA criterion. For the application to the canonical correlation criterion we had

3 Application: Landmark Specification in Water Temperature Data

to bootstrap a two dimensional time series, i.e. $z_{t,1}$ and $c_{t,1}$. The lag structure p for the residual autoregressive process was chosen by looking at the partial autocorrelation function which suggested to set $p = 3$ for the running means, temperature threshold and daily temperature curve, respectively. For the canonical correlation landmarks we set $p = 1$. For each day in our database we drew 5000 bootstrap replications and checked each time whether the landmark criteria were fulfilled. That is, for the k-th landmark we calculate the proportion of fulfillment around the observed values of the corresponding landmark criterion through

$$F_k\left(\tilde{t}\right) = \frac{1}{I \cdot 5000} \sum_{i=1}^{I} \sum_{j=1}^{5000} \mathbf{1}_{\{\tilde{L}^*_{i,k,j} \leq \tilde{L}_{i,k} - \tilde{t}\}}, \quad \tilde{t} = -20, \ldots, 20,$$

where $\tilde{L}^*_{i,k,j}$ is the j-th bootstrapped landmark in year i and $\mathbf{1}_{\{\cdot\}}$ is the indicator function. Note that $F_k(\cdot)$ can be interpreted as distribution function. The values were monotonically smoothed and are shown in Figure 3.9. Evidently, both running means landmarks show less variability than the remaining ones, followed by the temperature threshold landmarks exemplarily shown here for 11°C. Furthermore, for the first daily temperature curve (PCA) landmarks a high probability for negative deviations can be seen. Similarly, the second canonical correlation landmarks show less variability than the remaining two whose distribution functions can roughly be taken to be of the same shape.

In this section we demonstrated that landmarks defined through different criteria may also turn out to be of different variability. This may lead to inconsistencies among the landmarks within a year. However, we are not to concerned about that as we calculate time-warping functions by monotonically smoothing the data. This, in turn, allows us to handle inconsistencies automatically as we do not match the observed landmarks exactly to their reference points. We will now briefly sum up the results that can be drawn from this application.

3 Application: Landmark Specification in Water Temperature Data

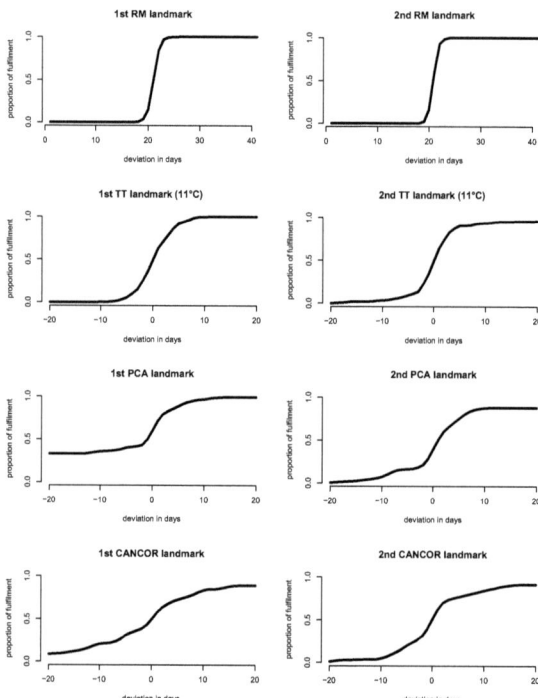

Figure 3.9: Monotonically smoothed proportions of fulfillment $F_k\left(\tilde{t}\right)$ of four landmark criteria applied to the bootstrap samples plotted against the deviation from the observed landmark $\tilde{t}$.

3.4 Results

In this chapter we suggested a modified landmark-based time-warping routine derived from the approach of Kneip & Gasser (1992) that can be applied to online-monitored time series data. We presented an application example for a water temperature dataset of the river Wupper. For these concrete data we proposed four different landmark criteria which led to up to twelve landmarks per year that can be calculated online. These observed landmarks were used to derive a time-warping function that allowed us to assess if the current year is running ahead or behind the average year. We tried to verify the time-warping using a spawn cycle dataset of the Brown Trout in the river Wupper. The ecological data matched our results in two of three years that were available. However, there is still not enough data available to assess the performance of our approach to the data example presented here. In a last step we provided insight in the variability of the landmarks derived from the different criteria.

4 Application: Forecasting Water Temperature with Dynamic Factor Models

As mentioned in Section 1.1 a forecast of water temperature at an upstream location is needed to determine the amount of waste heat that can be dissipated by the two powerplants in the city of Wuppertal using river water. This is only one example where such forecasts can be employed successfully. In general, water temperature is one of the steering influences that determine stream life and its forecasting is of key importance for the ecological management of waterbodies. Other authors focus on different aspects such as the impact of extreme weather conditions and global warming on the water temperature (see Morrill, Bales & Conklin, 2005).

Because of its importance there is a wide literature on the topic of water temperature forecasts. An overview of recent approaches can be found in Webb et al. (2008) or Caissie (2006). Although almost all articles identify the air temperature as a factor of great importance there are generally two types of strategies to produce forecasts. In some articles physical modelling techniques are pursued. For example, Caissie, Satish & El-Jabi (2005) use a heat exchange equation which contains the following variables: water temperature, daytime, distance downstream from the measurement site, mean water velocity, river width, dispersion coefficient in the direction of flow, specific heat of water, water density, wetted perimeter of the river, total heat flux from the atmosphere to the river and heat flux to or form the sediment or streambed. Many of these variables can hardly be measured, for instance, the heat flux from the atmosphere to the river

4 Application: Forecasting Water Temperature with Dynamic Factor Models

not only depends on air temperature but furhter decomposes into net long- and short-wave radiation, evaporative heat flux and convective heat transfer where the radiation depends on the cloud cover. In short, the physical approach is quite impractical as many of the influencing factors can only be estimated roughly which, in turn, leads to inaccurate forecasts. That is why in other articles (see Caissie, El-Jabi & Satish, 2001, for instance) a stochastic modelling approach is prefered considering the disturbances caused by inmeasureable factors as random and applying classical statistical tools instead. If forecasts shall be given on higher resolutions these models suffer from the *curse of dimensionality*, i. e. a strong increase in complexity if only a small number of dimensions is added. This, again, can lead to poor forecasts. That is, the reason why many studies from the hydrological field focus on predicting daily (or even weekly) mean or maximum temperatures only. A small collection of these modelling strategies with performance comparisons can be found in Caissie, El-Jabi & St-Hilaire (1998). As pointed out earlier in this thesis for our application example a high (hourly) resolution is essential as it is requiered to calculated an energy production limit and energy is traded in minute intervals. Two procedures to remedy the *curse of dimensionality* shall be compared within this chapter. The first strategy is to estimate an autoregressive model separately for each hour, that is, to allow not only the regression coefficients to vary over the day but also the set of influencing covariates. A similar approach using Bayesian methodology applied to an energy demand forecasting setup is pursued in Cottet & Smith (2003). A second and somewhat more elegant solution is the employment of dynamic factor models in VAR-form.

Dynamic factor models are applied in many different fields of study. Hyndman & Ullah (2007), for instance, suggest a DFM to forecast mortality and fertility rates and in Erbas, Hyndman & Gertig (2007) a similar model is used to predict breast cancer mortality rates. These are only examples from the broad literature where DFMs have been applied successfully in the past. An overview of the development stages of DFMs is given in Section 2.4.1. However, in our application we face the challenge to incorporate further candidate covariates. Water temperature is influenced by a variety of environmental variables. First of all it seems closely related to the air temperature and, furthermore, it

4 Application: Forecasting Water Temperature with Dynamic Factor Models

is also effected by global radiation, precipitation and stream flow. A first result from our analysis is that global radiation can be neglected because it does not show a significant effect as soon as air temperature is included. We also decided not to include precipitation and stream flow for the following reasons. Firstly, the effect of precipitation is partly contained in air temperature as a precondition for rain is cloud cover which results in less global radiation and, therefore, in lower air temperature, secondly, forecasts of precipitation are less reliable and also in many cases less available than air temperature forecasts and these will be needed in a forecasting setting. Thirdly, although rapid changes in stream flow caused by strong precipitation or manually by adjustments on one of the upstream dams in the river do have a significant impact, slower changes can be neglected. As explained in detail in Section 4.4 our dataset does contain rapid changes caused by human interference as well as by heavy rainfall but the human effects should be eliminated in new data measurements from the river Wupper as the *Wupperverband* agreed on a new water management policy that only allows slow manual changes of the stream flow. So the only problem that is left are rapid stream flow changes caused by heavy downpour and it is not our intention to find a forecasting procedure that is capable to handle extreme weather conditions. Hence, that special case is left aside in our analysis and we focus on the incorporation of lagged water temperature values as well as current and lagged air temperatures as covariates in our model. Note that the dependence of the response on the covariates is non-linear as we apply smooth principal components analysis (PCA) to estimate the relevant factors in the DFM.

Among others Caissie, Satish & El-Jabi (2005) point out that there are two different types of cycles contained in water temperature data. Firsly, the seasonal variation and, secondly, the daily variation which is stronger on summer days than in the winter. A feasible forecasting model should consider this interdependency. And the first step prior to apply any modelling techniques will be to remove this deterministic part from the data. The same is true for the air temperature. The analyses will then be carried out on the remaining residuals.

In Section 4.1 a method is presented that is apt to remove the deterministic component from the data and a detailed description of the different modelling approaches will be

4 Application: Forecasting Water Temperature with Dynamic Factor Models

given. Section 4.2 discusses how forecasts can be build and how their accuracy can be assessed and in Section 4.3 the results of the application of all approaches applied to our dataset will be presented. In Section 4.4 we will have a look on the data quality before Section 4.5 concludes.

4.1 Models and Estimation

In this section four competing approaches will be presented. In the first part it will be explained in detail how the deterministic component consisting of intra-day and seasonal cycles can be removed from the data accounting for their interdependencies. The modelling techniques described in the remainder of this section will be applied to the remaining component which is considered to be stochastic. In the second part a general formulation of the approximate dynamic factor model is given which allows for an influence of air temperature on the water temperature factors. Three strategies to compute the factor scores will be presented: firstly, a simple least squares approach, secondly, maximum likelihood based estimation for the water temperature scores only and, thirdly, maximum likelihood estimation for both, water and air temperature scores. This is followed by desription how model selection will be done as there are several parameters in the DFM that have to be calibrated. In the last part of this section we formulate an alternative approach that is used as benchmark. We propose an autoregressive model which is estimated for every hour of the day separately allowing parameters and the set of covariates to vary over the day.

4.1.1 Removing the Seasonal Component from the Data

Let $t = (i, d)$ be the time index consisting of year i and day of the year d where leap years will be ignored for simplicity, i.e. $d \in \{1, \ldots, 365\}$. We write $t+1$ for $(i, d+1)$ and if $t = (i, 365)$ we define $t+1 = (i+1, 1)$ and similarly for higher order differences at the end of a year. Note that in our application a year starts on 1st July and ends on 30th June. Our dataset consists of hourly temperature measurements and we concatenate the

4 Application: Forecasting Water Temperature with Dynamic Factor Models

values for day $t = (i, d)$ to a vector, that is, $\boldsymbol{w}_t = (w_{t,1}, \ldots, w_{t,24})^\top$ for water temperature and $\boldsymbol{a}_t$ is built analogously for the air temperature. Now the deterministic part shall be removed from the data. We therefore decompose both temperatures into

$$\boldsymbol{w}_t = \boldsymbol{\mu}_w(d) + \bar{\boldsymbol{w}}_t,$$
$$\boldsymbol{a}_t = \boldsymbol{\mu}_a(d) + \bar{\boldsymbol{a}}_t,$$

where $\boldsymbol{\mu}_w, \boldsymbol{\mu}_a : \{1, \ldots, 365\} \longrightarrow \mathbb{R}^{24}$ and can be interpreted as smooth functions giving the intra-day cycle on day d and over their entire domain the seasonal variation over the year. $\bar{\boldsymbol{w}}_t$ and $\bar{\boldsymbol{a}}_t$ are the 24-dimensional stochastic variations beyond the seasonal components which will be used for modelling.

As already pointed out the intra-day variation is stronger in summer than in winter and our definition of $\boldsymbol{\mu}_w$ and $\boldsymbol{\mu}_a$ is suited for handling this feature. Let $\bar{\boldsymbol{W}} = (\bar{\boldsymbol{w}}_1, \ldots, \bar{\boldsymbol{w}}_T)^\top$, and $\bar{\boldsymbol{A}}$ defined analogously, be $T \times 24$ matrices with daily water or air temperature measurements as row entries. We now estimate $\boldsymbol{\mu}_w$ and $\boldsymbol{\mu}_a$ with the help of a cyclic B-spline basis $\mathcal{B}_c(doy(t))$, see Section 2.1.4 or de Boor (1978) for details. Here $doy(t)$ denotes the day of the year corresponding to the t-th observation. Cyclic means that $\mathcal{B}_c(doy(t)) = \mathcal{B}_c(d_{t+365})$ holds which guarantees continuity at $\mu.(365)$ and $\mu.(1)$. To achieve a smooth fit without having to include a penalty term we choose only six knots per year where we place one knot at the 365th day of the year 365 (which is topologically identical with the zeroth day of the year) and put the remaining five knots equidistantly in between. Let $\mathcal{B}_c$ be the $T \times 6$ matrix with rows $(\mathcal{B}_c(doy(t)))^\top, t = 1, \ldots, T$ then we can estimate

$$\boldsymbol{\mu}_w = \mathcal{B}_c(\mathcal{B}_c^\top \mathcal{B}_c)^{-1} \mathcal{B}_c^\top \bar{\boldsymbol{W}} \quad \text{and} \quad \boldsymbol{\mu}_a = \mathcal{B}_c(\mathcal{B}_c^\top \mathcal{B}_c)^{-1} \mathcal{B}_c^\top \bar{\boldsymbol{A}},$$

where $\boldsymbol{\mu}_w$ and $\boldsymbol{\mu}_a$ are $T \times 24$ matices with rows $(\boldsymbol{\mu}_w(d))^\top$ and $(\boldsymbol{\mu}_a(d))^\top, d = 1, \ldots, T$, respectively. The temperature courses of two successive days, i.e. $\mu.(d)_{24}$ and $\mu.(d+1)_1$, can be connected smoothly as will be demonstrated in the application, see Section 4.3.

We now present the approximate dynamic factor model that will be used for our forecasting setting.

4 Application: Forecasting Water Temperature with Dynamic Factor Models

4.1.2 The Dynamic Factor Model

After having removed the deterministic component from the data we are left with the 24-dimensional residuals $\bar{w}_t$ and $\bar{a}_t$. Imposing a vectorautoregressive process directly on them would lead to a bad conditioned model because of the huge number of parameters to be estimated. Therefore, we will first extract a suitable number ($\ll 24$) of common factors form the stochastic component of both temperature types prior to apply any classical methods. That is, the residuals further decompose into

$$\bar{w}_t = \Lambda_w f_t + \epsilon_{w,t}, \qquad (4.1)$$

$$\bar{a}_t = \Lambda_a g_t + \epsilon_{a,t}, \qquad (4.2)$$

where f_t is a K-dimensional vector of water temperature factors, Λ_w is a $24 \times K$ dimensional loading matrix and $\epsilon_{w,t}$ as 24-dimensional white noise residual vector. Analogously, g_t is an H-dimensional vector of air temperature factor scores, Λ_a a $24 \times H$ dimensional matrix of factor loadings and $\epsilon_{a,t}$ the corresponding residual vector. Instead of using exploratory factor analysis the factors will be estimated using principal components analysis as this is more in line with famous approaches in the literature as Stock & Watson (2002a,b). A discussion of the difference between both techniques and under which conditions they yield approximately the same results can be found in Section 2.4.4. How the factor numbers K and H are fixed will be described in more depth in Section 4.1.2.2.

Let $\Delta_{a,b}, a < b$ denote the backshift operator defined by $\Delta_{a,b} f_t = (f_{t-a}^\top, \ldots, f_{t-b}^\top)^\top$. We now impose an autoregressive structure on the water temperature factors:

$$f_t = \underbrace{\beta_f}_{(K \times P_1 K)\text{-dim.}} \underbrace{(\Delta_{1,P_1} f_t)}_{(P_1 K \times 1)\text{-dim.}} + \underbrace{\beta_g}_{(K \times (P_2+1)H)\text{-dim.}} \underbrace{(\Delta_{0,P_2} g_t)}_{((P_2+1)H \times 1)\text{-dim.}} + \epsilon_{f,t}, \qquad (4.3)$$

with $\epsilon_{f,t}$ as K-dimensional white noise residual vector and β_f and β_g as coefficient matrices. Model 4.3 implies that today's water temperature factors depend on water temperature factors of the preceeding P_1 days and on air temperature factors of today and the preceeding P_2 days. If a forecast shall be made at timepoint t for timepoint $t + 1$ (or even further into the future) in a real forecasting setting the air temperature

4 Application: Forecasting Water Temperature with Dynamic Factor Models

of that day is unknown and has to be replaced by its meteorological forecast. However, for our forecast comparison study we use the observed temperatures (which in practise would be unknown) to avoid an increased amount of uncertainty due to the error in meteorological forecasts.

The common factors f_t and g_t in (4.3) are unobservable and have to be estimated. In the following section we will describe three routines of different complexity to approximate them.

4.1.2.1 Factor Estimation

The first approach is to use simple least squares estimation after having fixed the factor loadings. However, this disregards the stochastic models (4.1) – (4.3) and as a consequence the estimated parameters are not maximum likelihood based. We therefore propose two other strategies that involve simultaneos maximum likelihood estimation of the common factors and the parameters by applying an EM algorithm (see Section 2.2).

Least Squares Estimation (LS) The main advantage of this approach is its simplicity with the drawback that the estimated parameters β_f, β_g and the residual variances are not based on a maximum likelihood procedure and, therefore, may lack of some desired properties like asymptotical unbiasedness and consistency. The factor scores are simply taken as

$$\hat{f}_t = \Lambda_w^\top \bar{w}_t \quad \text{and} \quad \hat{g}_t = \Lambda_a^\top \bar{a}_t. \tag{4.4}$$

Given the factor scores, β_f and β_g can be estimated by applying ordinary least squares regression on equation (4.3).

Maximum Likelihood Estimation (ML) We now consider the stochastic models (4.1) and (4.3). That is, firstly, we assume that the residuals $\epsilon_{w,t}$ in (4.1) follow a normal distribution

$$\epsilon_{w,t} \sim \mathrm{N}(\mathbf{0}, \mathrm{diag}(\sigma_w^2)),$$

4 Application: Forecasting Water Temperature with Dynamic Factor Models

i. e., for simplicity we take the hourly variances to be independent. This is feasible as f_t and $\epsilon_{w,t}$ are independent by definition which leads to the decomposition

$$\operatorname{Var}(\bar{w}_t) = \Lambda_w \operatorname{Var}(f_t) \Lambda_w^\top + \operatorname{Var}(\epsilon_{w,t}), \tag{4.5}$$

with $\sigma_w^2 = (\sigma_{w,1}^2, \ldots, \sigma_{w,24}^2)$ and since Λ_w will be chosen to capture the biggest part of the variance, as described later, there is little information left in the last summand. For the residuals in equation (4.3) we assume normality, as well:

$$\epsilon_{f,t} \sim \mathrm{N}(\mathbf{0}, \operatorname{diag}(\sigma_f^2)),$$

with $\sigma_f^2 = (\sigma_{f,1}^2, \ldots, \sigma_{f,K}^2)$. An EM-algorithm (see Section 2.2) is applied to simultaneously fix the common factor scores f_t and to estimate the parameters β_f, β_g, σ_f^2 and σ_w^2. We will refer to the water temperature scores found by this method as $\hat{f}_t$. Note that for the air temperature factors we take the least squares estimates $\hat{g}_t$.

To simplify the formulation of the EM-algorithm we concatenate the parameters to a vector $\boldsymbol{\theta} = (\boldsymbol{\beta}_f^\top, \boldsymbol{\beta}_g^\top, (\sigma_f^2)^\top, (\sigma_w^2)^\top)^\top$ where the parameter matrices $\boldsymbol{\beta}_\cdot$ are stacked to vectors. Formally, the **E-Step** of the s-th iteration consists of the construction of the *Q-function*

$$Q(\boldsymbol{\theta}, \boldsymbol{\theta}^{(s-1)}) = \mathrm{E}_{\boldsymbol{\theta}^{(s-1)}}\big(l(\boldsymbol{\theta}; \bar{w}_t, f_t, g_t)\big),$$

where $l(\cdot)$ denotes the log-likelihood which, after dropping the constant term, is given by

$$l(\boldsymbol{\theta}; \bar{w}_t, f_t, g_t) = -\frac{1}{2} \sum_t \Bigg\{ \epsilon_{f,t} \operatorname{diag}(\sigma_f^2) \epsilon_{f,t}^\top + \sum_{k=1}^K \log(\sigma_{f,k}^2) \\ + \epsilon_{w,t} \operatorname{diag}(\sigma_w^2) \epsilon_{w,t}^\top + \sum_{j=1}^{24} \log(\sigma_{w,j}^2) \Bigg\}.$$

We denote the history at timepoint t with $H_t = (\Delta_{1,P_1} \hat{f}_t, \Delta_{0,P_2} \hat{g}_t)$. The only random components in the E-Step are the residuals which can be rewritten as $\epsilon_{f,t} = f_t - \beta_f(\Delta_{1,P_1} f_t) - \beta_g(\Delta_{0,P_2} g_t)$ and $\epsilon_{w,t} = \bar{w}_t - \Lambda_w f_t$. In order to determine the expected value of the log-likelihood function we have to calculate the conditional expectations

4 Application: Forecasting Water Temperature with Dynamic Factor Models

$\mathrm{E}(\boldsymbol{\epsilon}_{f,t}\mathrm{diag}(\boldsymbol{\sigma}_f^2)\boldsymbol{\epsilon}_{f,t}^\top|\bar{\boldsymbol{w}}_t,\boldsymbol{H}_t)$ and $\mathrm{E}(\boldsymbol{\epsilon}_{w,t}\mathrm{diag}(\boldsymbol{\sigma}_w^2)\boldsymbol{\epsilon}_{w,t}^\top|\bar{\boldsymbol{w}}_t,\boldsymbol{H}_t)$ for all t. For the former it suffices to compute $\mathrm{E}(\boldsymbol{f}_t\mathrm{diag}(\boldsymbol{\sigma}_f^2)\boldsymbol{f}_t^\top|\bar{\boldsymbol{w}}_t,\boldsymbol{H}_t)$ and $\mathrm{E}(\boldsymbol{f}_t|\bar{\boldsymbol{w}}_t,\boldsymbol{H}_t)$ as the remaining terms are known at timepoint t. Assume that we have calculated $\hat{\boldsymbol{f}}_{\tilde{t}} = \mathrm{E}(\boldsymbol{f}_{\tilde{t}}|\bar{\boldsymbol{w}},\boldsymbol{H}_t), \forall \tilde{t} \leq t-1$ we can compute the following two expectations which are unconditional with respect to $\bar{\boldsymbol{w}}_t$: $\check{\boldsymbol{f}}_t = \mathrm{E}(\boldsymbol{f}_t|\boldsymbol{H}_t) = \boldsymbol{\beta}_f(\Delta_{1,P_1}\hat{\boldsymbol{f}}_t) + \boldsymbol{\beta}_g(\Delta_{0,P_2}\hat{\boldsymbol{g}}_t)$ and $\check{\bar{\boldsymbol{w}}} = \mathrm{E}(\bar{\boldsymbol{w}}_t|\boldsymbol{H}_t) = \boldsymbol{\Lambda}_w\check{\boldsymbol{f}}_t$ where the latter can be defined as forecast of $\bar{\boldsymbol{w}}_t$ at timepoint $t-1$. We define

$$\begin{aligned}
\boldsymbol{\Sigma}_{ff} &= \mathrm{Var}(\boldsymbol{f}_t^\top|\boldsymbol{H}_t) &&= \mathrm{diag}(\boldsymbol{\sigma}_f^2), \\
\boldsymbol{\Sigma}_{\bar{w}\bar{w}} &= \mathrm{Var}(\bar{\boldsymbol{w}}_t^\top|\boldsymbol{H}_t) &&= \mathrm{diag}(\boldsymbol{\sigma}_w^2) + \boldsymbol{\Lambda}_w\boldsymbol{\Sigma}_{ff}\boldsymbol{\Lambda}_w^\top, \\
\boldsymbol{\Sigma}_{\bar{w}f} &= \mathrm{Cov}(\bar{\boldsymbol{w}}_t^\top,\boldsymbol{f}_t|\boldsymbol{H}_t) &&= \boldsymbol{\Lambda}_w\boldsymbol{\Sigma}_{ff}.
\end{aligned}$$

Following the standard results of the multivariate normal distribution the expected value of $\boldsymbol{f}_t$ conditional on $\bar{\boldsymbol{w}}_t$ is given by

$$\hat{\boldsymbol{f}}_t = \mathrm{E}(\boldsymbol{f}_t|\bar{\boldsymbol{w}}_t,\boldsymbol{H}_t) = \check{\boldsymbol{f}}_t + \boldsymbol{B}(\bar{\boldsymbol{w}}_t - \check{\bar{\boldsymbol{w}}}_t), \tag{4.6}$$

with $\boldsymbol{B} = (\boldsymbol{\Sigma}_{\bar{w}\bar{w}}^{-1}\boldsymbol{\Sigma}_{\bar{w}f})^\top$. Making use of the equivalence $\mathrm{Var}(X) = \mathrm{E}(X^2) - (\mathrm{E}(X))^2 \Leftrightarrow \mathrm{E}(X^2) = (\mathrm{E}(X))^2 + \mathrm{Var}(X)$ which is valid for any random variable X we get

$$\mathrm{E}(\boldsymbol{f}_t\mathrm{diag}(\boldsymbol{\sigma}_f^{-2})\boldsymbol{f}_t^\top|\bar{\boldsymbol{w}}_t,\boldsymbol{H}_t) = \hat{\boldsymbol{f}}_t\mathrm{diag}(\boldsymbol{\sigma}_f^{-2})\hat{\boldsymbol{f}}_t^\top + \mathrm{tr}\left[\mathrm{diag}(\boldsymbol{\sigma}_f^{-2})\mathrm{Var}(\boldsymbol{f}_t^\top|\bar{\boldsymbol{w}},\boldsymbol{H}_t)\right].$$

Using again standard results of the multivariate normal distribution the rightmost term on the right hand side can be rewritten as

$$\begin{aligned}
\mathrm{tr}\left[\mathrm{diag}(\boldsymbol{\sigma}_f^{-2})\mathrm{Var}(\boldsymbol{f}_t^\top|\bar{\boldsymbol{w}},\boldsymbol{H}_t)\right] &= \mathrm{tr}\left[\boldsymbol{\Sigma}_{ff}^{-1}(\boldsymbol{\Sigma}_{ff} - \boldsymbol{\Sigma}_{f\bar{w}}\boldsymbol{\Sigma}_{\bar{w}\bar{w}}^{-1}\boldsymbol{\Sigma}_{\bar{w}f})\right] \\
&= K - \mathrm{tr}(\boldsymbol{\Sigma}_{\bar{w}\bar{w}}^{-1}\boldsymbol{\Lambda}_w\boldsymbol{\Sigma}_{ff}\boldsymbol{\Lambda}_w^\top) \\
&= K - 24 + \mathrm{tr}(\boldsymbol{\Sigma}_{\bar{w}\bar{w}}^{-1}\mathrm{diag}(\boldsymbol{\sigma}_w^2)). \tag{4.7}
\end{aligned}$$

The number of principal components K will be chosen to cover the main part of the variance in $\bar{\boldsymbol{w}}$ which implies by equation (4.5) that the vector of the remaining variance not covered by the leading principal components, i.e. $\boldsymbol{\sigma}_w^2$, has relatively small entries and can therefore be neglected in (4.7). This leads to the approximation

$$\mathrm{E}(\boldsymbol{f}_t\mathrm{diag}(\boldsymbol{\sigma}_f^{-2})\boldsymbol{f}_t^\top|\bar{\boldsymbol{w}}_t,\boldsymbol{H}_t) \approx \hat{\boldsymbol{f}}_t\mathrm{diag}(\boldsymbol{\sigma}_f^{-2})\hat{\boldsymbol{f}}_t^\top + C_1,$$

4 Application: Forecasting Water Temperature with Dynamic Factor Models

and analogously to

$$\mathrm{E}(\boldsymbol{\epsilon}_{w,t}\mathrm{diag}(\boldsymbol{\sigma}_f^{-2})\boldsymbol{\epsilon}_{w,t}^\top|\bar{\boldsymbol{w}}_t, \boldsymbol{H}_t) \approx \hat{\hat{\boldsymbol{\epsilon}}}_{w,t}\mathrm{diag}(\boldsymbol{\sigma}_f^{-2})\hat{\hat{\boldsymbol{\epsilon}}}_{w,t}^\top + C_2,$$

where C_1 and C_2 are constants and $\hat{\hat{\boldsymbol{\epsilon}}}_{w,t} = \bar{\boldsymbol{w}}_t - \boldsymbol{\Lambda}_w\hat{\hat{\boldsymbol{f}}}_t$. Iterative calculation of these expected values completes the E-Step.

Once having built the *Q-function* the **M-Step** is easy as the likelihood function is maximized by the OLS estimates of the parameters using the expectation of the water temperature factors $\hat{\hat{\boldsymbol{f}}}_t^{(s)}$ in the s-th iteration.

As starting values $\hat{\hat{\boldsymbol{f}}}_t^{(0)}$ we take the LS-factors $\hat{\boldsymbol{f}}_t$ (see above) and iterate until $|\boldsymbol{\theta}^{(s)} - \boldsymbol{\theta}^{(s-1)}|$ is sufficiently small.

Full Maximum Likelihood Estimation (FullML) Up to this point we only made use of the LS air temperature factors but these are not based on a maximum likelihood estimation, either. In order to change this fact we extend the above idea by also incorporating a stochastic autoregressive model for the air temperature scores of the form

$$\boldsymbol{g}_t = \tilde{\boldsymbol{\beta}}_g(\Delta_{1,P_3}\boldsymbol{g}_t) + \boldsymbol{\epsilon}_{g,t}, \tag{4.8}$$

where we assume that the residuals are white noise, i.e.

$$\boldsymbol{\epsilon}_{g,t} \sim \mathrm{N}(\boldsymbol{0}, \mathrm{diag}(\tilde{\boldsymbol{\sigma}}_g^2)),$$

with $\tilde{\boldsymbol{\sigma}}_g^2 = (\tilde{\sigma}_{g,1}^2, \ldots, \tilde{\sigma}_{g,H}^2)$. For the residuals in (4.2) we assume

$$\boldsymbol{\epsilon}_{a,t} \sim \mathrm{N}(\boldsymbol{0}, \mathrm{diag}(\tilde{\boldsymbol{\sigma}}_a^2)).$$

We have to predict $\boldsymbol{g}_t$ based on $\boldsymbol{a}_1, \ldots, \boldsymbol{a}_t$, i.e., $\hat{\hat{\boldsymbol{g}}}_t = \mathrm{E}(\boldsymbol{g}_t|\boldsymbol{a}_t, \Delta_{1,\tilde{q}}\boldsymbol{g}_t)$ where we consider the current air temperature as known and in practice use a meteorological forecast. Figure 4.1 gives a graphical sketch of the dependence structure in a FullML-model for the lags $P_1 = 2$, $P_2 = 1$ and $P_3 = 2$. Once the expectation is estimated it can be inserted into the maximum likelihood routine of the ML approach. That is, to estimate the parameter vector $\tilde{\boldsymbol{\theta}} = (\boldsymbol{\theta}^\top, \tilde{\boldsymbol{\beta}}_g^\top, (\tilde{\boldsymbol{\sigma}}_g^2)^\top, (\tilde{\boldsymbol{\sigma}}_a^2)^\top)^\top$ we run a two stage EM-algoritm.

4 Application: Forecasting Water Temperature with Dynamic Factor Models

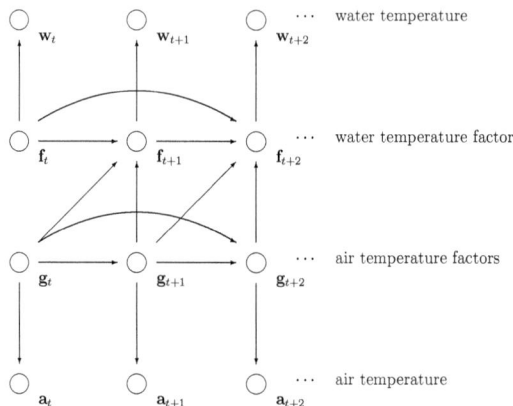

Figure 4.1: Graphical sketch of the dependence structure in the full factor model. Here the lags are set to $p=2$, $q=1$ and $\tilde{q}=2$.

The *Q-function* that has to be constructed in the **E-Step** of the s-th iteration is now given by

$$Q\big(\tilde{\boldsymbol{\theta}}, \tilde{\boldsymbol{\theta}}^{(s-1)}\big) = \mathrm{E}_{\tilde{\boldsymbol{\theta}}^{(s-1)}}\big(l(\tilde{\boldsymbol{\theta}}; \bar{\boldsymbol{w}}_t, \bar{\boldsymbol{a}}_t, \boldsymbol{f}_t, \boldsymbol{g}_t)\big),$$

4 Application: Forecasting Water Temperature with Dynamic Factor Models

and the log-likelihood additively expands to

$$l_{\text{full}}(\tilde{\boldsymbol{\theta}}; \bar{\boldsymbol{w}}_t, \bar{\boldsymbol{a}}_t, \boldsymbol{f}_t, \boldsymbol{g}_t) = l(\boldsymbol{\theta}; \bar{\boldsymbol{w}}_t, \boldsymbol{f}_t, \boldsymbol{g}_t)$$
$$- \frac{1}{2} \sum_t \left\{ \boldsymbol{\epsilon}_{g,t} \text{diag}(\tilde{\boldsymbol{\sigma}}_g^2) \boldsymbol{\epsilon}_{g,t}^\top + \sum_{h=1}^{H} \log(\tilde{\sigma}_{g,h}^2) \right.$$
$$\left. + \boldsymbol{\epsilon}_{a,t} \text{diag}(\tilde{\boldsymbol{\sigma}}_a^2) \boldsymbol{\epsilon}_{a,t}^\top + \sum_{j=1}^{24} \log(\tilde{\sigma}_{a,j}^2) \right\}.$$

where $\boldsymbol{\epsilon}_{a,t}$ and $\boldsymbol{\epsilon}_{g,t}$ are defined in (4.2) and (4.8), respectively. Let $\tilde{\boldsymbol{H}}_t = (\Delta_{1,P_3} \boldsymbol{g}_t)$ be the history for the air temperature factor scores. In complete analogy to the ML approach we have to estimate $\text{E}(\boldsymbol{g}_t|\bar{\boldsymbol{a}}_t, \tilde{\boldsymbol{H}}_t)$ and $\text{E}(\boldsymbol{g}_t \text{diag}(\tilde{\boldsymbol{\sigma}}_a^{-2}) \boldsymbol{g}_t^\top | \bar{\boldsymbol{a}}, \tilde{\boldsymbol{H}}_t)$ where we use the notation

$$\boldsymbol{\Sigma}_{gg} = \text{diag}(\tilde{\boldsymbol{\sigma}}_g^2), \quad \boldsymbol{\Sigma}_{\bar{a}\bar{a}} = \text{diag}(\tilde{\boldsymbol{\sigma}}_a^2) + \boldsymbol{\Lambda}_a \boldsymbol{\Sigma}_{gg} \boldsymbol{\Lambda}_a^\top \quad \text{and} \quad \boldsymbol{\Sigma}_{\bar{a}g} = \boldsymbol{\Lambda}_a \boldsymbol{\Sigma}_{gg}.$$

Following the argumentation given above we get

$$\hat{\tilde{\boldsymbol{g}}}_t = \text{E}(\boldsymbol{g}_t|\bar{\boldsymbol{a}}_t, \tilde{\boldsymbol{H}}_t) = \ddot{\boldsymbol{g}}_t + \tilde{\boldsymbol{B}}(\bar{\boldsymbol{a}}_t - \ddot{\bar{\boldsymbol{a}}}_t), \tag{4.9}$$

with $\tilde{\boldsymbol{B}} = (\boldsymbol{\Sigma}_{\bar{a}\bar{a}}^{-1} \boldsymbol{\Sigma}_{\bar{a}g})^\top$, $\ddot{\boldsymbol{g}}_t = \tilde{\boldsymbol{\beta}}_g(\Delta_{1,P_3} \hat{\tilde{\boldsymbol{g}}}_t)$ and $\ddot{\bar{\boldsymbol{a}}}_t = \text{E}(\bar{\boldsymbol{a}}_t|\tilde{\boldsymbol{H}}_t) = \boldsymbol{\Lambda}_a \ddot{\boldsymbol{g}}_t$. And as we choose the number of principal components for the air temperature h so that the main part of variance contained in the data is captured this leads to the following approximations:

$$\text{E}(\boldsymbol{g}_t \text{diag}(\tilde{\boldsymbol{\sigma}}_g^{-2}) \boldsymbol{g}_t^\top | \bar{\boldsymbol{a}}_t, \tilde{\boldsymbol{H}}_t) \approx \hat{\tilde{\boldsymbol{g}}}_t \text{diag}(\tilde{\boldsymbol{\sigma}}_g^{-2}) \hat{\tilde{\boldsymbol{g}}}_t^\top + C_3,$$
$$\text{E}(\boldsymbol{\epsilon}_{a,t} \text{diag}(\tilde{\boldsymbol{\sigma}}_a^{-2}) \boldsymbol{\epsilon}_{a,t}^\top | \bar{\boldsymbol{a}}_t, \tilde{\boldsymbol{H}}_t) \approx \hat{\tilde{\boldsymbol{\epsilon}}}_{a,t} \text{diag}(\tilde{\boldsymbol{\sigma}}_w^{-2}) \hat{\tilde{\boldsymbol{\epsilon}}}_{a,t}^\top + C_4,$$

where C_3 and C_4 are constants. Note that by using $\hat{\tilde{\boldsymbol{g}}}_t$ instead of $\hat{\boldsymbol{g}}_t$ in the history $\tilde{\boldsymbol{H}}_t$ defined above the prediction of $\boldsymbol{f}_t$ is effected, as well.

The **M-Step**, again, is easy as the *Q-function* is maximized by simply estimating all parameters by OLS regression.

Both steps are repeated until $|\boldsymbol{\theta}^{(s)} - \boldsymbol{\theta}^{(s-1)}|$ converges. As starting values the LS estimates $\hat{\boldsymbol{f}}_t$ and $\hat{\boldsymbol{g}}_t$ can be taken.

4.1.2.2 Model Selection

We need to select a suitable model for forecasting purposes and, therefore, a number of parameters has to be fixed. Firstly, we have to decide how many common factors for water and air temperatures are to be incorporated, that is, we have to choose K and H, respectively. Secondly, the number of timelags for both types of temperatures P_1 and P_2 have to be picked. In the FullML approach there is also the need to fix the timelag number P_3 for the air temperature model.

We split our dataset into two parts: a training sample which will be used to choose appropriate models for all three approaches (and a competing model that will be described later) and a forecasting sample where the model performances shall be compared. As we want to limit the numerical burden and to maintain interpretability we choose K and H to keep 99% of the total variation of the corresponding data. Furthermore, we set $P_3 = 2$, that is we assume the air temperature scores of the leading H common factors to follow a VAR(2) process or in other words the current air temperature course is presumed to depend only on the temperatures of the two preceeding days. This allows us to focus on the timelag selection in the approximate dynamic factor model (4.3). We therefore apply a multivariate Bayesian information criterion (BIC) to the estimated residuals $\hat{\epsilon}_{f,t}$ of equation (4.3) which we fit to our training data. For our application the BIC is given by

$$\text{BIC}_m(P_1, P_2) = \log(|\hat{\Sigma}_f|) + \frac{M(P_1, P_2)}{T} \log(T), \qquad (4.10)$$

where $|\hat{\Sigma}_f|$ is the determinant of the estimated covariance matrix of the residuals $\hat{\epsilon}_{f,t}$, T is the number of days in the training sample and the number of parameters in the model is given by $M(P_1, P_2) = K(P_1 k + (P_2 + 1)H)$. Optimal parameter combinations for all three dynamic factor models are chosen by minimizing (4.10) considering all possible combinations of $P_1 \in \{1, \ldots, 7\}$ and $P_2 \in \{0, \ldots, 7\}$.

In the last part of this section the autoregressive model will be introduced that will serve as benchmark in the data sample.

4 Application: Forecasting Water Temperature with Dynamic Factor Models

4.1.3 The Benchmark Model

Instead of fitting factor models we may consider the data $\bar{w}_t$ as simple univariate time series $\bar{w}_{\tilde{t}}$ with $\tilde{t} = (t, h)$ indicating year, day and hour of the day. This series can then be fitted with an autoregressive model applied directly to $\bar{w}_{\tilde{t}}$. It seems reasonable not only to allow that the autoregressive parameters vary over the hour of the day but to choose a model for each hour separately (see also Cottet & Smith, 2003). Assume that the water temperature depends on (a) the water temperature of the previous hours, (b) the water temperature at the same hour at previous days, (c) the air temperature at the same hour and day and possibly previous days and (d) the air temperature of the previous hours. To be specific, for $h = 1, \ldots, 24$ we assume

$$\begin{aligned}
\bar{w}_{t,h} &= \beta_{1,h}\bar{w}_{t-1,h} + \cdots + \beta_{L_1,h}\bar{w}_{t-L_1,h} \\
&+ \gamma_{1,h}\bar{w}_{t,h-1} + \cdots + \gamma_{L_2,h}\bar{w}_{t,h-L_2} \\
&+ \delta_{0,h}\bar{a}_{t,h} + \delta_{1,h}\bar{a}_{t-1,h} + \cdots + \delta_{L_3,h}\bar{a}_{t-L_3,h} \\
&+ \eta_{1,h}\bar{a}_{t,h-1} + \cdots + \eta_{L_4,h}\bar{a}_{t,h-L_4} + \epsilon_{t,h},
\end{aligned} \quad (4.11)$$

where $\bar{a}_{t,h}$ is the air temperature at time t and hour h. We have to choose the four lag lengths $L_1, \ldots, L_4$ and we therefore perform a hierarchical model selection in our training data starting by comparing all models which contain only one effect. As comparison criterion we use the ordinary (univariate) BIC given by

$$\text{BIC}_u(L_1, \ldots, L_4) = -2 \cdot l(L_1, \ldots, L_4; \bar{w}_t, \bar{a}_t) + M(L_1, \ldots, L_4) \cdot \log(T \cdot 24),$$

where $l(\cdot; \cdot)$ is the log-likelihood, $T \cdot 24$ is the number of observations (24 observations per day for T days in the training sample) and the number of parameters to be estimated is simply given by $M(L_1, \ldots, L_4) = L_1 + L_2 + (L_3 + 1) + L_4$. In each step we check if a smaller BIC_u value can be achieved by adding or dropping one term while preserving the hierarchy. We consider the model as benchmark since it follows more traditional time series modeling approaches. As seen below, however, the functional factor model from above performs better in the data example considered here.

4 Application: Forecasting Water Temperature with Dynamic Factor Models

In this section we described in detail the detrending of the data as well as the different types of models that shall be compared in this chapter. The more complicated estimation routines involving the EM-algorithm were presented and the model selection strategies have been explained. All steps mentioned here will be applied to the training sample. Having estimated the seasonal variations $\bar{w}_t$ and $\bar{a}_t$ and the optimal parameter combinations for all models we can move on to producing forecasts which will be the topic of the following section.

4.2 Forecasting

In the first part of this Section we will describe how forecasts for the next 24 hours can be produced. The second part is focused on longer forecasting horizons. The straight forward solution would be to forecast the first 24 hours and to plug the results in to produce a prediction for the following day and so on. However, the results of this procedure would lack in continuity between the single days and an approach will be presented that can remedy this deficit. In the third part three criteria to measure the forecasting performance will be introduced. Each one focuses on a different feature and, later, in Section 4.3 all models will be compared in all three criteria. In the last part of this section we present two methods to estimate pointwise forecasting intervals. The first way to do this is to define intervals of constant width for all days of the year. Although this approach is simple and may be justified for some applications a more realistic one is to assume the width of the forecasting intervals to vary over the seasons as the data shows strong water temperature variation during the warming up phase in spring and only little changes in winter, for example.

4.2.1 One Day Ahead Forecast

Now, let $t = (i, d)$ be a day of the forecasting sample and we want to give a forecast for day $t + 1 = (i, d + 1)$, i.e. the next 24 hours. To do so, we need observations of

4 Application: Forecasting Water Temperature with Dynamic Factor Models

water and air temperature up to timepoint t and additionally the air temperature at timepoint $t+1$ which, in practise, is unknown and has to be replaced by a meteorogical forecast. However, in order to avoid forecasting errors caused by the uncertainty of these forecasts we will use the observed temperatures in our comparison study. Assume that all historical data up to timepoint $t-1$ is known and at timepoint t there arrives new information: the courses of water and air temperatures of that day and the forecasted air temperatures for the next 24 hours $\boldsymbol{a}_{t+1}$. For the Least Squares approach we calculate $\hat{\boldsymbol{f}}_t$ and $\hat{\boldsymbol{g}}_{t+1}$ based on equation (4.4):

$$\hat{\boldsymbol{f}}_t = \hat{\boldsymbol{\Lambda}}_w^\top \bar{\boldsymbol{w}}_t \quad \text{and} \quad \hat{\boldsymbol{g}}_{t+1} = \hat{\boldsymbol{\Lambda}}_a^\top \bar{\boldsymbol{a}}_{t+1},$$

where $\bar{\boldsymbol{a}}_{t+1} = \boldsymbol{a}_{t+1} - \hat{\boldsymbol{\mu}}_a(d+1)$. For the Maximum Likelihood method we take the LS estimate $\hat{\boldsymbol{g}}_{t+1}$ and run an E-Step to obtain $\hat{\hat{\boldsymbol{f}}}_t$ based on equation (4.6). For the Full Maximum Likelihood approach a second E-Step following (4.9) is needed to get $\hat{\hat{\boldsymbol{g}}}_{t+1}$. Plugging in the new information into equation (4.3) yields a water temperature factor forecast

$$\dot{\boldsymbol{f}}_{t+1} = \boldsymbol{\beta}_f(\Delta_{1,P_1} \tilde{\boldsymbol{f}}_{t+1}) + \boldsymbol{\beta}_g(\Delta_{0,P_2} \tilde{\boldsymbol{g}}_{t+1}),$$

where we set $\tilde{\boldsymbol{f}}_t = \hat{\boldsymbol{f}}_t$ and $\tilde{\boldsymbol{g}}_t = \hat{\boldsymbol{g}}_t$ for LS, $\tilde{\boldsymbol{f}}_t = \hat{\hat{\boldsymbol{f}}}_t$ and $\tilde{\boldsymbol{g}}_t = \hat{\boldsymbol{g}}_t$ for ML and $\tilde{\boldsymbol{f}}_t = \hat{\hat{\boldsymbol{f}}}_t$ and $\tilde{\boldsymbol{g}}_t = \hat{\hat{\boldsymbol{g}}}_t$ for FullML. We can then define the water temperature forecast by

$$\dot{\boldsymbol{w}}_{t+1} = \hat{\boldsymbol{\mu}}_w(d+1) + \hat{\boldsymbol{\Lambda}}_a \dot{\boldsymbol{f}}_{t+1}.$$

Note that for the benchmark model the construction of forecasts is straight forward. We simply have to iteratively forecast the next hour to obtain a one (or m days) ahead forecast.

4.2.2 Longer Forecasting Horizons

Meteorological air temperature forecasts are nowadays given for a quite impressive long forecasting horizon. Our intention is to make use of this information and develop a multiple day ahead forecast for the water temperature as well. In principle, one could easily plug in the one day ahead forecast $\dot{\boldsymbol{w}}_{t+1}$ together with the meteorological air

4 Application: Forecasting Water Temperature with Dynamic Factor Models

temperature forecast $\dot{a}_{t+2|t}$ for day $t+2$ issued at day t in our model. This allows to obtain $\dot{w}_{t+2|t}$, the forecast for day $t+2$ issued at day t, and so on. However, our approach does not guarantee that our forecast is continuous in the sense that $\dot{w}_{i(d+1)24}$ and $\dot{w}_{i(d+2)1}$ might not connect. To correct for this deficit there are in principle two possibilities. Firstly, after calculation of the forecasted values one could use a simple smoothing step to connect the forecasts. Alternatively, we could restructure the model by binning data to time intervals of length m days. Let therefore index $t = (i, \tilde{d})$ stand for year i and day sequence $\tilde{d} = (d, d+1, \ldots, d+m-1)$. Hence, $\boldsymbol{w}_t$ is an $(m \cdot 24)$-dimensional vector and $t+1 = (i, d+m, \ldots, d+2m-1)$. We can now run the same modelling exercise as above but with higher dimensional time series for water and air temperature. Note that dependent on the starting day for the binning we get m different models. In practice, we use all m models and take the average of the resulting m estimates to run our forecast.

4.2.3 Forecasting Performance

After model selection and estimation have been carried out on the training sample, that is after having fixed P_1 and P_2 for each of the dynamic factor models and $L_1, \ldots, L_4$ for the benchmark model, the forecasting performance is measured in the evaluation sample. We therefore make use of the prediction error. Let $\dot{\boldsymbol{w}}_t$ denote the 24-dimensional forecasted water temperature vector at timepoint t obtained by one of the above mentioned dynamic factor models or the benchmark model. The prediction error is expressed with the following measurements

Mean Squared Prediction Error $\quad \text{MSPE} = \frac{1}{T'} \sum_{t=1}^{T'} (\boldsymbol{w}_t - \dot{\boldsymbol{w}}_t)^\top (\boldsymbol{w}_t - \dot{\boldsymbol{w}}_t),$

Mean Maximum Prediction Error $\quad \text{MMPE} = \frac{1}{T'} \sum_{t=1}^{T'} \max |\boldsymbol{w}_t - \dot{\boldsymbol{w}}_t|,$

Mean Squared Prediction Error for the Maximum $\quad \text{MSPM} = \frac{1}{T'} \sum_{t=1}^{T'} \left(\max(\boldsymbol{w}_t) - \max(\dot{\boldsymbol{w}}_t) \right)^2,$

where T' is the number of days in the forecasting sample. Note that each error criterion focuses on a different feature. MSPE gives the mean of the accumulated hourly

4 Application: Forecasting Water Temperature with Dynamic Factor Models

forecasting errors while MMPE measures the mean maximum daily difference between hourly forecast and observed temperature. Finally, MSPM is used to evaluate the mean difference between the daily forecasted and observed maximum temperature and it will be used in Section 4.3.3 to allow a comparisons of our models to different approaches suggested in the literature of hydrology which only focus on forecasting the daily maximum temperature.

4.2.4 Forecasting Errors

Constant Forecasting Intervals The three fitting strategies (LS, ML and FullML) allow directly the calculation of forecasting intervals in the following form. Taking the LS estimates let $\hat{\Sigma}_{\epsilon_w}$ be the estimated covariance matrix based on the fitted model residuals $\hat{\epsilon}_{w,t} = \bar{w}_t - \Lambda_w \hat{f}_t$. Accordingly, let $\hat{\Sigma}_{\epsilon_f}$ be the estimated covariance matrix of $\epsilon_{f,t}$ based on the fitted dynamic factor model (4.3). Assuming $\epsilon_{w,t}$ and $\epsilon_{f,t}$ are independent it follows directly that

$$\text{Var}(\dot{w}_{t+1}) = \Lambda_w \Sigma_{\epsilon_f} \Lambda_w^\top + \Sigma_{\epsilon_w}.$$

Assuming normality, a (pointwise) 95% forecasting interval is then obtained roughly by

$$\dot{w}_{t+1} \pm 1.96 \sqrt{\text{diag}\left(\widehat{\text{Var}}(\dot{w}_{t+1})\right)}. \qquad (4.12)$$

We are ignoring estimation variability here, which is justifiable given the amount of data we have at hand. For ML and FullML the forecasting intervals are calculated analogously but with their fitted residuals resulting in different estimates for the variance matrices Σ_{ϵ_w} and Σ_{ϵ_f}, respectively. Constant forecasting intervals, however, ignore the changes of the variance over the year which can be observed in the dataset and are therefore a suboptimal but easy solution if forecasting accuracy shall be assessed.

Time Varying Forecasting Intervals In order to account for the heteroscedasticity over the seasons the estimated covariance matrices can be considered as functions in the day of the year d, i.e. $\hat{\Sigma}_{\epsilon_w}(d)$ and $\hat{\Sigma}_{\epsilon_f}(d)$. In our application we use the weights of the

4 Application: Forecasting Water Temperature with Dynamic Factor Models

Epanechnikov kernel which are defined by

$$\mathcal{K}(x) = \begin{cases} \frac{3}{4}(1-x^2), & \text{if } x \in [-1;1] \\ 0, & \text{if } x \notin [-1;1], \end{cases}$$

and calculate the first covariance matrix estimate in the following way:

$$\hat{\boldsymbol{\Sigma}}_{\epsilon_w}(d) = \frac{\sum_{t=1}^{T} \mathcal{K}\left(\frac{d-doy(t)}{b}\right) \hat{\boldsymbol{\epsilon}}_{w,t} \hat{\boldsymbol{\epsilon}}_{w,t}^\top}{\sum_{t=1}^{T} \mathcal{K}\left(\frac{d-doy(t)}{b}\right)},$$

where b is a bandwidth and $doy(t)$ is the day of the year for the t-th observation. $\hat{\boldsymbol{\Sigma}}_{\epsilon_f}(d)$ can be computed analogously. We replace the variance term in equation (4.12) with

$$\widehat{\text{Var}}(\dot{\boldsymbol{w}}_{t+1}, d) = \boldsymbol{\Lambda}_w \hat{\boldsymbol{\Sigma}}_{\epsilon_f}(d) \boldsymbol{\Lambda}_w^\top + \hat{\boldsymbol{\Sigma}}_{\epsilon_w}(d).$$

In our application this will lead to wider forecasting intervals in spring when the river water is warming up rapidly and in narrower ones in winter. This allows to assess the quality of a forecast more reliably. Note that there is no straight forward extension of this technique to m-days-ahead forecasts. For those we will give only constant forecasting intervals.

In this section a detailed description of the construction of one-day ahead and smoothly connected m-days ahead forecasts was given. We introduced three error criteria, each one focusing on a different feature what makes it usefull for varying purposes. Finally, we offered two possibilities to construct forecasting intervals which enable us to assess the accuracy of a forecast. These methods will be applied to the evaluation sample and the results are presented in the following section.

4.3 Application to the Dataset

Our data range from 1 July 2002 to 30 June 2007. As mentioned earlier we let a year start in July and end in June so that we have hourly observations over five years of

4 Application: Forecasting Water Temperature with Dynamic Factor Models

data in total. We use the first four years as training sample and will do forecasting performance comparisons on the evaluation sample which ranges from 1 July 2006 to 30 June 2007.

In the first part of this section we describe the results of our DFM-forecasts in detail before we will compare it with the benchmark model in the second part. In the third part we will make use of the MSPM criterion introduced earlier to compare the DFM's with some approaches taken from the literature on hydrology which only aim to forecast daily maximum water temperatures.

4.3.1 Forecasting Water Temperature with Dynamic Factor Models

At first we estimate the smooth seasonal components $\boldsymbol{\mu}_w(d)$ and $\boldsymbol{\mu}_a(d)$ for water and air temperature from the training sample as described in Section 4.1.1. Figure 4.2 illustrates the results. The left column corresponds to water and the right to air temperature. In each column the topmost panel shows in black colour the observed temperature curves of the four years of the estimation sample at 16:00 which is, in the mean, the warmest hour of the day refering to water temperature. The red curve is the respective smooth function, i.e. the 16th column of $\boldsymbol{\mu}_w$ or $\boldsymbol{\mu}_a$, respectively. The green curve is the smooth function for 07:00 which is the coldes hour in the mean (refering to water temperature). Evidently, the difference between both smooths is bigger in summer than in winter. Below the vertical line the cyclic basis functions $\mathcal{B}_c$ are plotted. The mid panel shows the entire dataset including estimation and forecasting sample, separated by a vertical blue dashed line. In red the smooth component is plotted this time all 24 hours stacked. Note, that for the forecasting sample the smooth $\boldsymbol{\mu}.(d)$ has been extrapolated. The bottom panel shows a "zoom in" of the former on the 5 to 10 May 2007. It illustrates that the discrete μ-functions can be interpreted as smooth terms. Note that, if interpreted as continous functions, they do not show discontinuities between sequential days but connect smoothly.

In a second step we perform a model selection using the multivariate BIC, see Section 4.1.2.2, for all three DFM's applied to the evaluation sample. As pointed out earlier we have to choose the number of factors for water and air temperature which we fixed

4 Application: Forecasting Water Temperature with Dynamic Factor Models

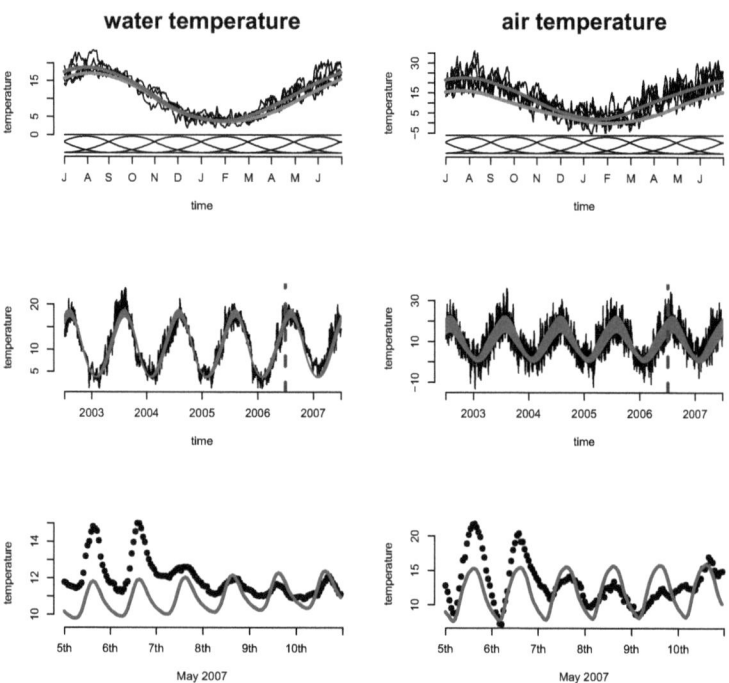

Figure 4.2: Plots in the left column show water temperatures, in the right one air temperatures. Top panel: Temperature curves of four years at 16:00 with corresponding smooth (red) and smoothed curve for 07:00 (green). Cyclic B-spline basis functions are shown below. Mid panel: Temperatures of estimation and forecasting sample (divided by a blue dashed line) with 24-hour smooth. Bottom panel: "Zoom in" on the panel above for the time span 5 to 10 May 2007.

4 Application: Forecasting Water Temperature with Dynamic Factor Models

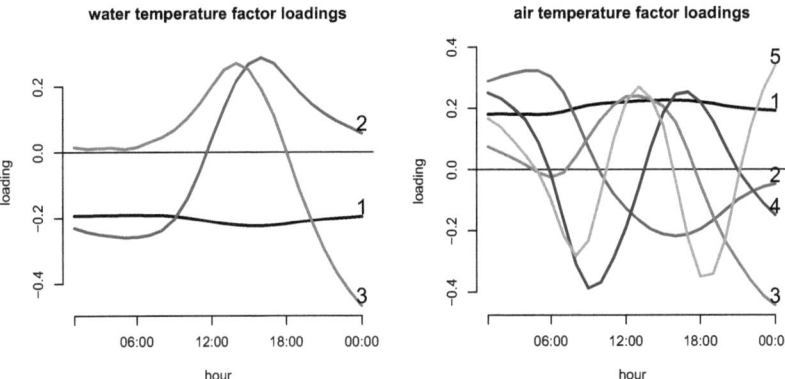

Figure 4.3: Left panel: Loadings of the first three water temperature factors for the LS method. Right panel: Loadings of the first five air temperature factors for the LS method.

to $K = 3$ and $H = 5$, respectively. The factors are estimated by a principal components analysis and our parameter choice maintains more than 99% of the total variation in the data for all three approaches. The first three LS factors are shown in the left panel of Figure 4.3 and cover 94.5%, 4.1% and 0.7%, respectively. While the first factor captures the overall level, the second and third mirror the daily variation. The first five LS factors for the air temperatures are shown in the right panel, they cover 83.8%, 10.8%, 2.8%, 1.0% and 0.6%, respectively. Note that the eigenvectors are not cyclic and do not start and end at the same point. This is, in fact, desirable since the eigenvectors in this way also capture changes in the (short term) level of the temperature, i.e. increments or decrements of the temperature over the days.

Furthermore, we fixed the lag for the stochastic model of the air temperature (4.8) in the FullML approach at $P_3 = 2$ so we can focus on the parameter selection of P_1

4 Application: Forecasting Water Temperature with Dynamic Factor Models

Fitting	optimal setting		criterion		
approach	P_1	P_2	MSPE	MMPE	MSPM
LS	3	1	2.334	0.483	0.162
ML	3	1	2.383	0.486	0.165
FullML	3	1	2.516	0.502	0.173
Benchmark	—	—	3.636	0.596	0.252
Model (4.13)	—	—	—	—	1.545
Model (4.14)	—	—	—	—	4.507
Model (4.15)	—	—	—	—	0.970

Table 4.1: Performance comparison of all modelling approaches described in this chapter. The upper part gives the optimal lag values for the factor approaches estimated by minimizing the multivariate BIC.

and P_2 in model (4.3). After finishing the model selection we produced one-day-ahead forecasts for the entire evaluation sample. The results of all approaches together with the optimal parameter combinations are shown in Table 4.1. We see that the LS approach yields the smallest prediction error in all three criteria followed by the ML and FullML approach. This might be due to the fact, that we considered the factor loadings as fixed and estimated these by a PCA; which by itself ignores serial correlation as does the LS estimation. Overall, the performance of the model is adequate for forecasting as will be seen in the remaining part of this section. For simplification we will make use of the LS method from now on.

The forecasting errors for the one-day-ahead LS forecast are visualized in the top panels of Figure 4.4. Overall these errors are smaller in winter months than in summer and are biggest for May and June when water starts to warm up quickly. Exemplarily, we take two time windows to demonstrate the forecasting properties. First, we look at the days 30 July to 6 August 2006. The mid panel of Figure 4.4 shows the data together with an one-day-ahead LS forecast. The corresponding 95% forecasting intervals were calculated using an Epanechnikov kernel and a bandwidth $b = 45$. Apparently, the

4 Application: Forecasting Water Temperature with Dynamic Factor Models

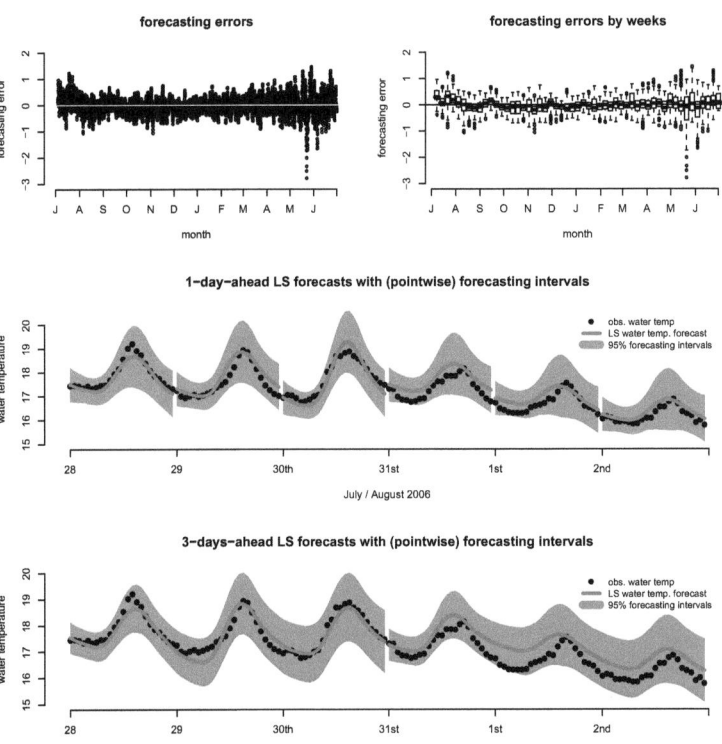

Figure 4.4: Left top panel: Hourly LS forecasting errors $\dot{w}_t - w_t$. Right top panel: Box-plots of hourly forecasting errors grouped by calendar weeks. Mid panel: One-day-ahead LS forecasts with pointwise time varying forecasting intervals. Bottom panel: Three-days-ahead LS forecasts with pointwise constant forecasting intervals.

4 Application: Forecasting Water Temperature with Dynamic Factor Models

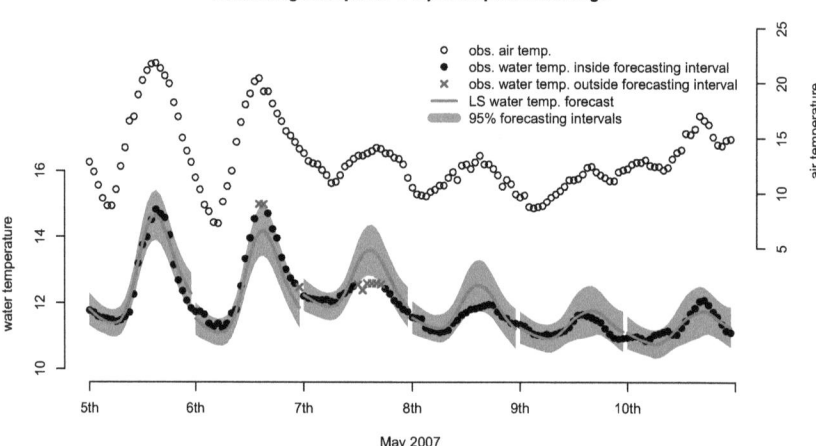

Figure 4.5: White circles indicate air temperature observations. Water temperature observations are shown as solid circles when lying in the forecasting intervals and as red crosses otherwise. The green line gives the one-day-ahead LS forecast for the water temperature and the grey areas mark the corresponding 95% forecasting intervals.

forecast behaves satisfactory as it suitably fits the data but we do see the discontinuities at the break of the days. This can be corrected using a three-days-ahead forecast, which is shown in the bottom panel of Figure 4.4 calculated at 29th July for 30th July to 1st August, and then calculated at 1st August 1st for 2nd to 4th August. As pointed out earlier there is no straight forward extension of time varying forecasting intervals for three days ahead forecasts and therefore constant intervals are given. Obviously, they are quite narrow for the short range forecast and broaden for the second and third forecasted day which is a consequence of the increasing uncertainty. Note that we do not

4 Application: Forecasting Water Temperature with Dynamic Factor Models

see any observation outside the 95% forecasting intervals in both panels although there are 144 hourly observations. This fact is due to the data quality and will be commented later in Section 4.4. Secondly, in Figure 4.5 we take up the zoom-in example shown in the bottom panels of Figure 4.2. This gives an example how a rapid change of the air temperature is handled by the model, here, again, fitted with the LS method. The one-day-ahead forecasts have been performed for 5 to 10 May 2007. On 7th May a significant drop of the air temperature was observed. That day the maximum air temperature was about 7°C lower than on the preceeding day. The LS forecasting method is not able to describe this rapid change directly and some consecutive observations drop below their corresponding forecasting intervals. Those measurements are indicated as crosses in the plot. On the following day, however, the situation has been adapted to and the temperature lies within the forecasting intervals, again. Overall, the performance of our forecast is promising. Daily variation is exhibited and does change gradually over the year, dependent on the autoregression. Temperature changes are captured and the model looks capable to be applied in real practise.

4.3.2 Comparison to the Benchmark Model

The benchmark model was estimated for each hour separately as described in Section 4.1.3. It can be seen in Table 4.2 that there are only few hours where water temperatures of the previous day (at the same hour) show a significant influence. More important are water temperatures of preceeding hours. For most hours there are two to five earlier measurements needed but in one case (hour 18) one includes up to 8 lagged hours. For almost all timepoints the air temperature at the same hour was found significant. Only few models contain air temperatures of previous days. Dependencies of water temperature on foregoing air temperature measurements were important mostly for daylight hours.

We compare the benchmark model to the dynamic factor models using again the forecasting sample from 1 July 2006 to 30 June 2007. The model performance is included in Table 4.1. The dynamic factor models show clear advantages in all criteria. They also perform well compared to other modelling approaches form the hydorlogical field

4 Application: Forecasting Water Temperature with Dynamic Factor Models

hour	1	2	3	4	5	6	7	8	9	10	11	12
L_1	1	–	–	–	1	–	–	–	–	–	1	1
L_2	4	4	4	2	3	2	3	4	4	2	2	4
L_3	0	0	0	0	1	0	0	–	0	0	0	0
L_4	–	–	–	3	–	–	–	3	4	4	5	2
hour	13	14	15	16	17	18	19	20	21	22	23	24
L_1	1	2	–	–	–	1	1	1	–	–	–	1
L_2	5	5	2	3	6	8	2	3	4	5	4	4
L_3	0	0	0	0	0	2	0	0	0	0	0	0
L_4	3	3	3	3	3	1	1	1	1	2	–	–

Table 4.2: Optimal lag lengths for the different hourly benchmark models (4.11) chosen by the univariate BIC.

as can be seen in the next section.

4.3.3 Comparison to Other Modellling Approaches in Hydrology

To the best of our knowledge there is no earlier attempt to forecast water temperatures on an hourly basis so that no direct competitors are available. There are, however, alternative approaches which focus on forecasting the daily maximum temperature. Using our approach we can look at the forecasted daily maximum temperature and we obtain a forecasting accuracy of order $\sqrt{\text{MSPM}} \approx 0.403°C$ (The benchmark model yields $\sqrt{\text{MSPM}} \approx 1.072°C$). This value is now compared with three approaches listed in Caissie, El-Jabi & St-Hilaire (1998). In analogy to our paper they extract a seasonal component to guarantee a first-order stationarity of the water and air temperature time series, respectively. They suggest to fit the first harmonic component of a Fourier series to the data or, alternatively, a sinusoidal function and to apply the modelling analysis on the remaining air and water temperature residuals. We follow their idea but use a cyclic B-spline basis. With $t = (i, d)$ as above the decompositon of water and air temperature

4 Application: Forecasting Water Temperature with Dynamic Factor Models

maximum has the form $w_t^{\max} = \mu_w^{\max}(d) + \bar{w}_t^{\max}$ and $a_t^{\max} = \mu_a^{\max}(d) + \bar{a}_t^{\max}$, where $\mu_w^{\max}$ and $\mu_a^{\max}$ are the seasonal components for maximum water and air temperature, respectively, and $\bar{w}_t^{\max}$ and $\bar{a}_t^{\max}$ give the corresponding temperature residuals. In analogy to our analysis the first four years are used to estimate the parameters and the fifth year with the same index set J as used above is the evaluation sample.

The first model presented in Caissie, El-Jabi & St-Hilaire (1998) is a multiple regression analysis of the water temperature on the current and lagged air temperature (Kothandaraman, 1971). Using our notation the model has the form

$$\bar{w}_t^{\max} = (\Delta_{0,2}\bar{a}_t^{\max})\boldsymbol{\beta}_1 + \epsilon_t^{(1)}, \qquad (4.13)$$

where $\boldsymbol{\beta}_1$ is a three dimensional parameter vector and $\epsilon_t^{(1)}$ are the residuals at time point t. Performing a least squares regression and fitting the model yields $\sqrt{\text{MSPM}} = 1.243°\text{C}$. The second model is a second-order Markov process which was first suggested by Cluis (1972) and takes into account the autoregressive structure of the water temperature data, that is,

$$\bar{w}_t^{\max} = (\Delta_{1,2}\bar{w}_t^{\max})\boldsymbol{\beta}_2 + \mathcal{T}\bar{a}_t^{\max} + \epsilon_t^{(2)}, \qquad (4.14)$$

with $\boldsymbol{\beta}_2 = \left(\rho_1(1-\rho_2)/(1-\rho_1^2), (\rho_2-\rho_1^2)/(1-\rho_1^2)\right)^\top$ and ρ_1 and ρ_2 as the autocorrelation coefficients of the water temperature maximum for a lag of 1 and 2, respectively. In (4.14), $\mathcal{T}$ is the thermal exchange coefficient which can be estimated by least squares after determining $\boldsymbol{\beta}_2$ and substracting the corresponding terms from equation (4.14). Applied to our dataset this model yields $\sqrt{\text{MSPM}} = 2.123°\text{C}$. The third model presented in Caissie, El-Jabi & St-Hilaire (1998) follows a Box-Jenkins approach (see Box, Jenkins & Reinsel, 1994) applied to an equation described in Marceau, Cluis & Morin (1986), i.e.,

$$\begin{aligned}\bar{w}_t^{\max} &= (\delta_1 + \phi_1)\bar{w}_{t-1}^{\max} - \delta_1\phi_1\bar{w}_{t-2}^{\max} + \zeta_0\bar{a}_t^{\max} \\ &\quad - \zeta_0\phi_1\bar{a}_{t-1}^{\max} + n_t - \delta_1 n_{t-1} + \epsilon_t^{(3)},\end{aligned} \qquad (4.15)$$

with parameters estimated minimizing the squared residual sum. The prediction error was calculated to $\sqrt{\text{MSPM}} = 0.985°\text{C}$. Table 4.1 sumarizes the performance results of all mentioned models.

4 Application: Forecasting Water Temperature with Dynamic Factor Models

Note that for the estimation and for the forecasting of the DFM's and the benchmark model we used 23 times more data than was used for the three approaches described in this section. Therefore, it is not surprising that they are all clearly beaten by the former described models. However, we did this comparison to emphasize the great advantage that bring our models over those more classical methods from the hydrology field. We also want to point out that hourly data can be obtained at low cost. Hence, the application of our more complex models can easily be justified.

In this section the models described in this chapter were applied to a dataset of hourly water temperature measurements of the river Wupper. It turned out that the LS method of the dynamic factor models is the overal best performing approach. Competitors like the autoregressive benchmark model or univariate approaches are outperformed easily. In the next section we will shortly comment on the quality of the dataset at hand.

4.4 Data Quality

The mid and bottom panels in Figure 4.4 display a forecasting example and give the corresponding pointwise 95% forecasting intervals. Six days of data are shown which results in a total of 144 hourly observations. If the intervals are estimated correctly one should expect to find at least some measurements lying outside. As this is not the case the conclusion can be drawn that the intervals are too wide. Figure 4.6 shows whether an observation of the forecasting sample is inside or outside its corresponding forecasting interval and the red line indicates the smoothed coverage probability which evidently varies over the year. Overall, we achieved a 95.3% coverage as we desired but for some months it is 100% while in the spring time when water is heating up it sinks under 90% and in May even under 80%. We observed this effect when we first calculated intervals of constant width over the year and tried to counter it by letting the width vary in time as described in Section 4.2.4. We therefore experimented with

4 Application: Forecasting Water Temperature with Dynamic Factor Models

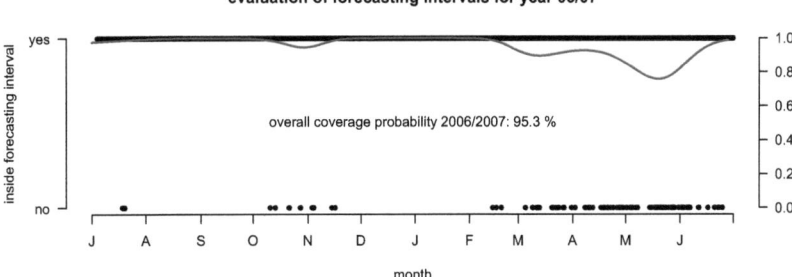

Figure 4.6: Black dots indicate whether an observation was inside or outside its forecasting interval. The red line gives the smoothed coverage probability. The global coverage probability was 95.3%.

different bandwidths b and it turned out that, although small enhancements could be achieved, the coverage probability could not be evened out. The result can be seen in Figure 4.6. We see the main error source for the forecasting interval width calculation in the data quality. Firstly, there seem to have been a lot of missing values which were simply interpolated linearly. This created a lot of artificial outliers. The top panel of Figure 4.7 displays an example of multiple periods which were obviously interpolated linearly. Similar data errors can be found in both training and forecasting sample and it would have been more helpful if missing data were simply flagged as *Not Available* so that we could decide whether to plug in an appropriate daily temperature course or to drop the respective observations from all calculations. Secondly, along with the data came stream flow measurements which we did not and do not want to account for. It is beyond question that it has a significant impact on the heat up process of stream water but as pointed out earlier upstream of the measurement site Laaken there are a number of dams and the municipal utility agreed on another water management policy to avoid abrupt level changes in the future. Nevertheless, such changes are contained in our data

4 Application: Forecasting Water Temperature with Dynamic Factor Models

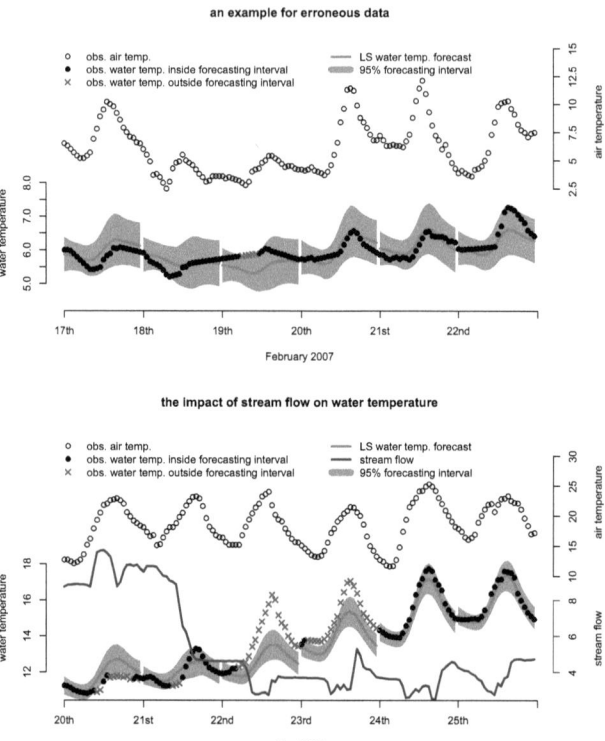

Figure 4.7: Top panel: Observed water and air temperature from 17 to 22 February 2007 with corresponding point and interval forecasts. A part of the water temperature observations seem to be a result of linear interpolation. Bottom panel: Observed water and air temperature and stream flow measurements from 20 to 25 May 2007 with corresponding point and interval forecasts.

4 Application: Forecasting Water Temperature with Dynamic Factor Models

and as can be seen in the bottom panel of Figure 4.7 our forecasting model is unable to handle these situations. In the example the stream was at a high level on Sunday, 20 May 2007 and after having started work on Monday the water supervisor closed a dam which reduced the level from 10-12m^3/sec to around 4m^3/sec. As a consequence the water temperature started to rise drastically although the air temperature stayed at a more or less constant level and our model needs two days to adopt to the new setting. However, as situations like this are improbable in future data we think disregarding stream flow is feasible to keep the model as simple as possible.

4.5 Results

In this chapter we demonstrated the use of dynamic factor models for forecasting water temperature. DFMs can be used to compress high-resolution temperature measurements so that the dimension of the original time series to be forecasted is significantly reduced without loosing too much information. The dimension reduction allows to employ vectorautoregressive time series models for the forecasting equation.

Our dynamic factor models easily outperform univariate models from the field of hydrology which were built to forecast the daily maximum temperature. This is unsurprising as our models are able to handle the large amount of data that comes with hourly measurements. However, this is a key feature of DFMs. The additional data necessary can be obtained at low cost and its incorporation yields a huge performance improvement.

We also used another classical autoregressive model as benchmark for our DFMs which was outperformed, as well. Furthermore, we compared three different estimation routines for the latent common factors. The outcome of this is that the computationally easiest approach yielded the best results which is an advantage for the practical implementation.

The LS routine described in this chapter has already been implemented and the tool is employed by the Wuppertal municipal utility.

5 Application: Forecasting Energy Demand with Dynamic Factor Models

In this chapter we will present a second application of dynamic factor models. We will develop a general approach for the forecasting of energy demand. This is also partly motivated through the need of water temperature management described in Section 1.1. However, besides of being universally applicable to energy demand forecasting settings our approach also performs well compared to established forecasting models in this field. We will demonstrate the abilities of our methodology considering two data examples. The first concerns the district heating demand in a citywide steam network in the city of Wuppertal in Germany, the second gives the electricity demand in the state of Victoria, Australia.

Forecasting demand for energy at an intraday resolution is an important problem for both energy utilities and management organisations for a number of reasons. The first is to ensure short term system stability; for example, to maintain a voltage range across an electricity grid. The second is for infrastructure planning and maintenance; for example, to ensure essential network maintenance is undertaken during times of low demand, or using longer term forecasts to decide the mix of electricity generation capacity or natural gas storage facilities that are built. The third reason is due to the introduction of wholesale electricity and natural gas markets in many regions. In wholesale markets suppliers and distributors bid for energy prior to dispatch, resulting in a spot price for the commodity that varies intraday. These markets can be highly

5 Application: Forecasting Energy Demand with Dynamic Factor Models

volatile, and participants rely on accurate forecasts of intraday demand to pursue optimal bidding strategies. Another example is the framework of dynamic water temperature management in the river Wupper, see Section 1.1. Here, a forecast of the hot tap water demand is requiered to calculate the amount of waste heat that has to be dissipated into the river. In this chapter we therefore treat energy demand in general and propose new methodology to forecast energy demand at an intraday resolution. We employ recent developments in semiparametric regression methodology to capture calendar and meteorological components. We focus on forecast horizons of up to one week, and show that residual serial dependence can be captured using multivariate time series models which greatly enhance forecast accuracy.

The literature on intraday energy demand forecasting has been dominated by the forecasting of electricity demand (also called load), although many of the models and methods proposed are equally applicable to natural gas and steam. The literature is extensive, and so is the number of competing models. For horizons of one day or less univariate time series models (Taylor, Menezes & McSharry, 2006; Taylor & McSharry, 2007; Taylor, 2010) and artificial neural networks (Park et al. 1991; Darbellay and Slama 2000) have proven particularly popular. Weather effects can be included, although there is little evidence in the literature that there is much improvement to be made at very short horizons of 6 hours of less. For longer horizons, semiparametric regression methods have also been used to capture the extensive periodic component of demand, along with weather based effects; for example, see Harvey & Koopman (1993), Smith (2000), Martín-Rodríguez & Cáceres-Hernández (2005) and Panagiotelis & Smith (2008b). Multivariate regression and time series models with dimension equal to the number of intraday periods have proven successful for forecasting demand over horizons of up to one week; see Ramanathan et al. (1997), Cottet & Smith (2003), Soares & Medeiros (2008) and references therein. And many studies have looked at the issue of how temperature affects intraday demand in different locations (Harvey and Koopman 1993; Pardo, Meneu and Valor 2002; Liu et al. 2006; Cancelo, Espasa and Grafe 2008; Hor, Watson and Majithia 2005), and occasionally other meteorological variables (Cottet & Smith, 2003; Panagiotelis & Smith, 2008b). Dordonnat et al. (2008) go a

5 Application: Forecasting Energy Demand with Dynamic Factor Models

different route and model the effect of covariates as a time-varying parameter model, which is similar to the time-varying periodic spline methodology of Harvey & Koopman (1993). Recent overviews on modeling and forecasting electricity demand at an intraday resolution can be found in Weron (2006), Soares & Medeiros (2008) and Taylor (2010).

Though the literature on electricity demand forecasting is well developed this does not apply in the same way to the forecasting of heating demand. We refer to Nielsen & Madsen (2000, 2006) and Dotzauer (2002) as central contributions in this field. We will demonstrate in the chapter that both tasks, heating as well as electricity demand forecasting can be handled with the same statistical model.

Our approach combines many of the insights and characteristics of previous approaches. We use semiparameteric regression methodology to estimate marginal mean demand using both calendar (time of year and day type) and meteorological variables. We introduce flexible high-dimensional basis representations for the unknown functions, but ensuring smoothness by introducing a shrinkage penalty for the basis terms. Such an approach has proven popular in the statistical literature (Eilers & Marx, 1996; Ruppert, Wand & Carroll, 2003; Wood, 2006) and technical details are summarized in Section 2.1. It has the advantage of allowing for semiparametric interaction effects (which are apparent in our problem), and is both numerically quite stable and computationally efficient for complex models; see Ruppert, Wand & Carroll (2009) for a comprehensive survey and discussion. We use a multivariate model, where each intraday period has a separate mean model. Allowing for diurnal variation in model components is a common feature of a number of successful forecasting models (Cottet & Smith, 2003; Soares & Medeiros, 2008).

We also consider two multivariate time series models for residual demand that exploit the fact that the multivariate vector is longitudinal (i.e., that the elements of the multivariate vector are contiguous intraday observations). The first of these is a periodic autoregression (PAR) (Pagano, 1978; Franses & Paap, 2004), which can account for seasonal or periodic structure in a time series. This is the case here with period equal to one day, and PARs have been used to forecast intraday electricity prices previously (Broszkiewicz-Suwaja et al. 2004; Guthrie and Videbeck 2007; Panagiotelis and Smith

5 Application: Forecasting Energy Demand with Dynamic Factor Models

2008a). The second multivariate time series model we consider is an approximate dynamic factor model (Sargent & Sims, 1977; Geweke, 1977). These are used extensively to forecast macroeconomic time series; see Breitung & Eickmeier (2006), Stock & Watson (2006) and Hallin & Liška (2007) for recent overviews and a detailed description of the methodology as well as a literature review can be found in Section 2.4. Dynamic factor models decompose a multivariate time series into a component driven by a low-dimensional dynamic latent factor, and an orthogonal idiosyncratic error. Estimation of the factors and factor loadings can be carried out with principal component analysis (PCA) applied to the covariance matrix of the process (see Stock & Watson 2002a, 2002b) or the spectral density matrix (see Forni, Hallin, Lippi & Reichlin 2000, 2004, 2005). A third approach is to use functional PCA, see Ramsay & Silverman (2005). We apply all three estimation methods and compare the resulting forecasts with those obtained using the PAR.

We consider the following data examples. The first is forecasting demand for district heating in a citywide steam network in the German city of Wuppertal. Steam for the network is supplied from co-generating power stations, and is used for both space and water heating. Co-generating power stations are energy efficient, have a low carbon footprint, and have high market penetration in northern and central Europe, as well as areas of North America. We find a strong temperature effect for heating demand that is well captured using air temperature, along with minimum and maximum daily temperature, but as interactions with the season. Over-and-above calendar and weather effects, including a multivariate time series component dramatically improves the accuracy of forecasts by between 54% one day ahead, and 34% seven days ahead. The second application is to aggregate electricity demand in the state of Victoria, Australia. Here, weather sensitivity of demand proves to be strong, largely due to the high variation in meteorological conditions. Again, including a multivariate time series component substantially improves the forecasts over-and-above calendar and weather effects. Over the period January to September 2009, these improvements have an average daily monetary value in the wholesale market of between $39,976 one day ahead and $9,040 seven days ahead. In both applications, we find the dynamic factor models to provide better forecasts than the popular PARs.

The chapter is structured as follows. In Section 5.1 we outline the PAR and the three variations on the dynamic factor models we employ to forecast residual demand. In Section 5.2 we outline the semiparametric regression methodology we use to estimate the marginal mean, based on calendar and meteorological variables. Sections 5.3 and 5.4 apply the methodology to the German district heating and Victorian electricity problems, respectively; Section 5.5 concludes.

5.1 Time Series Component

Let y_t denote energy demand observed at times $t = 1, \ldots, T$. For each observation, denote the intraday period as $h(t)$, the day in the sample as $d(t)$, and the day of the year as $doy(t)$. We assume the data are observed at N equally-spaced time points during a day, so that $1 \leq h(t) \leq N$. In our first application the data are observed hourly, so that $N = 24$, while in the second the data are observed every half hour, so that $N = 48$. We assume that y_t depends upon covariates $\boldsymbol{x}_t$, which in our analysis includes calender and meteorological variables, the influence of which differs during the hours of the day. We employ the regression model

$$y_t = \mu_{h(t)}(\boldsymbol{x}_t) + u_t, \quad t = 1, \ldots, T, \tag{5.1}$$

where $\mu_{h(t)}$ is the regression function at hour $h(t)$, and u_t is residual demand with $E(u_t) = 0$. The specification of $\mu_{h(t)}(\boldsymbol{x}_t)$ is discussed in Section 5.2, while the component u_t is assumed to be a stationary time series, for which we consider two models: a periodic autoregression (PAR), and a dynamic factor model.

5.1.1 Periodic Autoregression

If $h = h(t)$, the PAR model is

$$u_t = \sum_{j \in \mathcal{L}_h} u_{t-j} \beta_{h,j} + \sigma_h z_t, \tag{5.2}$$

where $\beta_{h,j}$ are the autoregressive coefficients, σ_h^2 the variances, z_t is a white noise process and $\mathcal{L}_h$ a set of integers representing the lag structure in period h. We set

$$\mathcal{L}_h = \{1, \ldots, L_{h,1}, N, 2N, \ldots, L_{h,2}N\},$$

so that an observation at intraday period h is directly related to the $L_{h,1} < N$ observations immediately prior, and the observations at the same intraday period on the $L_{h,2}$ preceding days. Following Franses & Paap (2004; Chapter 3), equation (5.2) can be rewritten in vector form as follows. Let $d = d(t)$, $\boldsymbol{u}_{(d)} = (u_{N(d-1)+1}, \ldots, u_{Nd})^\top$, then

$$\boldsymbol{\Psi}_0 \boldsymbol{u}_{(d)} = \sum_{j=1}^{P} \boldsymbol{\Psi}_j \boldsymbol{u}_{(d-j)} + \boldsymbol{z}_{(d)},$$

where $\boldsymbol{z}_{(d)} = (z_{N(d-1)+1}, \ldots, z_{Nd})^\top$, $P = 1 + \lfloor (\max_h(\mathcal{L}_h) - 1)/N \rfloor$ and $\boldsymbol{\Psi}_0, \ldots, \boldsymbol{\Psi}_P$ are $(N \times N)$ matrices with non-zero elements

$$\boldsymbol{\Psi}_0[h, k] = \begin{cases} 1 & \text{if } h = k, \\ 0 & \text{if } h < k, \\ -\beta_{h,h-k} & \text{if } h > k, \end{cases}$$

$$\boldsymbol{\Psi}_j[h, k] = \beta_{h, h+Nj-k},$$

with $\beta_{h,j} = 0$ if $j \notin \mathcal{L}_h$. Multiplying on the left by $\boldsymbol{\Psi}_0^{-1}$ results in the 'reduced form' vector autoregression

$$\boldsymbol{u}_{(d)} = \sum_{j=1}^{P} \boldsymbol{\Psi}_0^{-1} \boldsymbol{\Psi}_j \boldsymbol{u}_{(d-j)} + \boldsymbol{e}_{(d)}, \qquad (5.3)$$

with sparse coefficient matrices $(\boldsymbol{\Psi}_0^{-1} \boldsymbol{\Psi}_j)$, $j = 1, \ldots, P$ and

$$\operatorname{Var}(\boldsymbol{e}_{(d)}) = \boldsymbol{\Psi}_0^{-1} \operatorname{diag}(\sigma_1, \ldots, \sigma_N)(\boldsymbol{\Psi}_0^{-1})^\top = \boldsymbol{\Sigma}_e.$$

Considering the form of $\boldsymbol{\Psi}_0$, the precision matrix $\boldsymbol{\Sigma}_e^{-1}$ is band diagonal with $\max_h(L_{h,1})$ bands.

Estimation can be undertaken using the reduced form at equation (5.3) as suggested by Pagano (1978). This approach is pursued by Panagiotelis & Smith (2008a), who assume $L_{h,1} = 2$ and $L_{h,2} = 1$, so that $P = 1$. However, they assume the matrix $\Psi_0^{-1}\Psi_1$ is well-approximated using a sparse matrix with nonzero elements along the leading diagonal and the three elements in the upper right hand corner. In our work we assume $z_t \sim N(0,1)$, and estimate the unknown parameters by maximising the likelihood constructed directly from equation (5.2). We condition on pre-period observations, so that the likelihood is separable in the parameters broken down by hour h. In this case, maximising the likelihood with respect to $\{\beta_{h,j}, \sigma_h; j \in \mathcal{L}_h\}$ separately for each hour corresponds to full maximum likelihood estimation. Stationarity can be checked (or enforced) using the reduced form at equation (5.3) as discussed in Reinsel (1993) for a vector autoregression. In both our empirical applications we find the estimated PAR to be stationary.

The orders $L_{h,1} \geq 0$ and $L_{h,2} \geq 0$ are chosen by stepwise model selection based on the BIC. Again, because the likelihood is separable we can undertake this for each hour h separately. A step in the stepwise algorithm considers up to four moves from a current position $(L_{1,h}, L_{2,h})$ corresponding to adding $(-1,0)$, $(0,-1)$, $(1,0)$ or $(0,1)$ to the position, and ignoring inadmissable values. Whichever of the (up to) five positions that results in the lowest BIC is selected at that step. The initial position of this algorithm is $(L_{1,h} = 1, L_{2,h} = 0)$ and steps repeated until no move is made.

After estimation, point forecasts of future values of u_t, for $t > T$, can be evaluated directly from equation (5.2).

5.1.2 Dynamic Factor Models

In our second time series model we assume $\boldsymbol{u}_{(d)}$ results from a stochastic process which is driven by $M \ll N$ unobserved latent factors $\boldsymbol{f}_{(d)} = (f_{d1}, \ldots, f_{dM})^\top \in \mathbb{R}^M$, so that

$$\boldsymbol{u}_{(d)} = \Lambda \boldsymbol{f}_{(d)} + \boldsymbol{\epsilon}_{(d)} = \boldsymbol{\eta}_{(d)} + \boldsymbol{\epsilon}_{(d)}, \tag{5.4}$$

where $\Lambda \in \mathbb{R}^{N \times M}$ is a matrix of orthonormal factor loadings and $\boldsymbol{\epsilon}_{(d)} = (\epsilon_{d1}, \ldots, \epsilon_{dN})^\top$ is a vector of white noise disturbances. Typically, M is much less than N, thereby reducing the dimension of the problem greatly; for example, $M = 3$ in our model of heat demand and $M = 4$ in our model of electricity demand. To identify the first two moments of $\boldsymbol{u}_{(d)}$

5 Application: Forecasting Energy Demand with Dynamic Factor Models

we assume that $E(\boldsymbol{f}_{(d)}) = \boldsymbol{0}$, $E(\boldsymbol{\epsilon}_{(d)}) = \boldsymbol{0}$, $E(\boldsymbol{f}_{(d)}\boldsymbol{\epsilon}_{(d)}^\top) = \boldsymbol{0}$ and $E(\boldsymbol{\epsilon}_{(d)}\boldsymbol{\epsilon}_{(d)}^\top) = \boldsymbol{\Sigma}_\epsilon$. The factors are also assumed to be orthogonal, so that $E(\boldsymbol{f}_{(d)}\boldsymbol{f}_{(d)}^\top) = \boldsymbol{\Omega} = \mathrm{diag}(\omega_1, \ldots, \omega_m)$, and

$$\mathrm{Var}(\boldsymbol{u}_{(d)}) = \boldsymbol{\Sigma}_u = \boldsymbol{\Lambda}\boldsymbol{\Omega}\boldsymbol{\Lambda}^\top + \boldsymbol{\Sigma}_\epsilon = \boldsymbol{\Sigma}_\eta + \boldsymbol{\Sigma}_\epsilon. \tag{5.5}$$

In the dynamic model the factors also follow a vector autoregression over days d of the form

$$\boldsymbol{f}_{(d)}|\boldsymbol{f}_{(d-1)}, \ldots, \boldsymbol{f}_{(d-L)} = \sum_{l=1}^{L} \boldsymbol{B}_l \boldsymbol{f}_{(d-l)} + \boldsymbol{w}_{(d)}. \tag{5.6}$$

Here the $\boldsymbol{B}_1, \ldots, \boldsymbol{B}_L$ are ($M \times M$) autoregressive coefficient matrices and $\boldsymbol{w}_{(d)}$ is a zero mean disturbance with $\mathrm{Cov}(\boldsymbol{w}_{(d)})$ a diagonal matrix.

Dynamic factor models are extensively treated in Section 2.4 and we outline below three different ways to obtain estimates of both the factors and loading matrix $\boldsymbol{\Lambda}$ mainly following Stock & Watson (2002a, 2002b) or Forni, Hallin, Lippi & Reichlin (2000, 2004, 2005). Conditional on the factor estimates, the unknown parameters in equation (5.6) can be estimated using maximum likelihood with the assumption that $\boldsymbol{w}_{(d)}$ follows a Gaussian distribution. The lag length L is obtained by minimising a multivariate BIC over the values $\{0, 1, \ldots, 10\}$. We check stationarity of equation (5.6) as outlined in Reinsel (1993, p. 26), and find that all our empirical results correspond to stationary processes.

Method 1: Principal Components Least Squares

The first method is an approximate, but simple, approach based on a principal component analysis (PCA) applied to the empirical covariance of $\boldsymbol{u}_{(d)}$, denoted by $\hat{\boldsymbol{\Sigma}}_u$. This is in line with the approach of Stock & Watson described in Section 2.4.2. Let $\tilde{\boldsymbol{\Lambda}}$ denote the ($N \times M$) dimensional matrix of the M orthogonal eigenvectors of $\hat{\boldsymbol{\Sigma}}_u$ corresponding to the M largest eigenvalues. Predicted factors are then obtained by a least squares estimate denoted by

$$\tilde{\boldsymbol{f}}_{(d)}^{LS} = (\tilde{\boldsymbol{\Lambda}}^\top \tilde{\boldsymbol{\Lambda}})^{-1} \tilde{\boldsymbol{\Lambda}}^\top \boldsymbol{u}_{(d)} = \tilde{\boldsymbol{\Lambda}}^\top \boldsymbol{u}_{(d)}.$$

Stock & Watson (2002a) show this provides consistent estimates of the factors if both $N, T \to \infty$. However, for N fixed, which is the case in our intraday demand models,

5 Application: Forecasting Energy Demand with Dynamic Factor Models

the estimates are inconsistent unless $\mathbf{\Sigma}_\epsilon = \sigma_\epsilon^2 \mathbf{I}_N$. Nevertheless, this is likely to have only a minor impact in our analysis because we select M so that the columns of $\tilde{\mathbf{\Lambda}}$ explain in excess of 90% of the variation in $\mathbf{u}_{(d)}$, and $\mathbf{\Sigma}_\eta$ dominates $\mathbf{\Sigma}_\epsilon$ in the variance decomposition at equation (5.5). A detailed justification for this is given in Section 2.4.4.

Method 2: Principal Components Functional Data Analysis

The second approach results from noting that the elements of $\mathbf{u}_{(d)}$ are likely to be serially dependent. This motivates an assumption that the columns of $\mathbf{\Lambda}$ are smooth functions of the hour h, which we denote as $\mathbf{\Lambda}(h) = [\mathbf{\Lambda}_1(h), \ldots, \mathbf{\Lambda}_m(h)]$, for $h = 1, \ldots, N$. Considering $\mathbf{u}_{(d)}$ as a random function $\mathbf{u}_{(d)}(h)$ evaluated at the N intraday periods corresponds to functional data. The analysis of such data is called 'functional data analysis"; for example, see Ramsay & Silverman (2005) and Ferraty & Vieu (2006). Estimation can be carried out with functional principal component analysis, which corresponds to PCA with an additional smoothing penalty on the estimate of $\mathbf{\Lambda}_j(h)$; see Ramsay & Silverman (2005, Chapter 9.3). Denoting the resulting first m fitted functional principal components with $\tilde{\mathbf{\Lambda}}(h)$, least squares predictions $\bar{\mathbf{f}}_{(t)}^{LS} = \tilde{\mathbf{\Lambda}}(h)^\top \mathbf{u}_{(t)}(h)$ for the factors can be computed.

Functional principal component analysis has been applied and investigated in a number of scenarios; see for example Kneip & Utikal (2001) or Cardot, Faivre & Goulard (2003). Theoretical properties are based on the Karhunen-Loève representation (see Loève, 1978):

$$\mathbf{u}_{(d)}(h) = \sum_{q=1}^{\infty} F_q \mathbf{\Lambda}_q(h), \qquad (5.7)$$

where $\mathbf{\Lambda}_q(h)$ is the q-th column of $\mathbf{\Lambda}(h)$, $F_q = \int \mathbf{u}_{(d)}^\top(h) \mathbf{\Lambda}_q(h) dh$ are random variables with $\mathrm{E}(F_q) = 0$, $\mathrm{Var}(F_q) = \lambda_q$ and $\lambda_1 \geq \lambda_2 \geq \lambda_3 \geq \ldots \geq 0$ are the eigenvalues of the covariance function $\mathrm{E}\big(\mathbf{u}_{(d)}(h)\mathbf{u}_{(d)}(h)^\top\big)$. Assuming that $\lambda_q = 0$ for all $q > M$ allows consistent estimation if $N, T \to \infty$, as shown in Chiou, Müller & Wang (2003). However, when N is fixed, the estimates are approximate but not necessarily consistent. Again, we are unconcerned about this because we select the M eigenfunctions to account for in

5 Application: Forecasting Energy Demand with Dynamic Factor Models

excess of 90% of the variation, so that

$$\frac{\sum_{q=1}^{M} \lambda_q}{\sum_{q=M+1}^{N} \lambda_q} > 9. \tag{5.8}$$

Method 3: Spectral Density Decomposition

The third approach follows Forni, Hallin, Lippi & Reichlin (2000, 2004, 2005) who propose to base the principal component analysis not on the (empirical) covariance matrix $\hat{\Sigma}_u$ of $\boldsymbol{u}_{(d)}$, but on the spectral density matrix, see Section 2.4.3. Defining the matrix of order k autocorrelations as $\boldsymbol{\Gamma}_k = E\left(\boldsymbol{u}_{(d)}\boldsymbol{u}_{(d-k)}^\top\right)$, the spectral density matrix is

$$\boldsymbol{\Phi}_u(\theta) = \frac{1}{2\pi} \sum_{k=-\infty}^{\infty} e^{-ik\theta} \boldsymbol{\Gamma}_k,$$

for $\theta \in [-\pi, \pi]$. In practice, the summation is approximated with a finite sum, and in our empirical work we compute from $k = -30$ to $k = 30$, which corresponds to a time lag of one month; see Forni, Hallin, Lippi & Reichlin (2005). Applying PCA to $\boldsymbol{\Phi}_u(\theta)$ results in the following decomposition for the spectral density matrix:

$$\boldsymbol{\Phi}_u(\theta) = \boldsymbol{R}(\theta)\boldsymbol{P}(\theta)\boldsymbol{R}(\theta) + \boldsymbol{\Phi}_\epsilon(\theta).$$

Here, $\boldsymbol{R}(\theta) \in \mathbb{C}^{N \times k}$ is the matrix of the eigenvectors corresponding to the k largest eigenvalues $\rho_1(\theta) \geq \ldots \geq \rho_k(\theta)$, $P(\theta) = \text{diag}(\rho_1(\theta), \ldots, \rho_k(\theta))$ and $\boldsymbol{\Phi}_\epsilon(\theta)$ is the part of the spectral density matrix that remains unexplained. Using this decomposition, the variation of $\boldsymbol{u}_{(d)}$ can be divided into two components:

$$\text{Var}\left(\boldsymbol{u}_{(d)}\right) = \int_{-\pi}^{\pi} \boldsymbol{R}(\theta)\boldsymbol{P}(\theta)\boldsymbol{R}^\top(\theta)\mathrm{d}\theta + \int_{-\pi}^{\pi} \boldsymbol{\Phi}_\epsilon(\theta)\mathrm{d}\theta = \check{\boldsymbol{\Sigma}}_\eta + \check{\boldsymbol{\Sigma}}_\epsilon, \tag{5.9}$$

where the first integral can be evaluated numerically in practise. We estimate the elements of the factors $\boldsymbol{f}_{(d)}$ as $\check{f}_{dj} = \check{\boldsymbol{a}}_j^T \boldsymbol{u}_{(d)}$, where $\check{\boldsymbol{a}}_j$ is the generalized eigenvector resulting from a generalized eigenvalue decompostion of the couple $\check{\boldsymbol{\Sigma}}_\eta$ and $\check{\boldsymbol{\Sigma}}_\epsilon$ that corresponds to the j-th largest generalized eigenvalue. Details and an explicit description is found in Forni et al. (2005) and are therefore omitted here. Forni et al. (2005) show in a number of simulations that their method provides better forecasts than the simple PCA combined with least squares estimation of the factors outlined above. However, we show that this does not extend to the demand forecasts in our empirical work.

All three methods require the estimation of the vector autoregression in equation (5.6), but employ different estimates of the factors $\boldsymbol{f}_{(d)}$ and the loading matrix $\boldsymbol{\Lambda}$ in equation (5.4). Using these, future values can be forecasted from the dynamic factor models as

$$\hat{\boldsymbol{u}}_{(d)} = \sum_{l=1}^{L} \hat{\boldsymbol{\Lambda}} \hat{\boldsymbol{B}}_l \hat{\boldsymbol{f}}_{(d-l)}, \quad \text{for } d > d_0. \tag{5.10}$$

Here, $\{\hat{\boldsymbol{f}}_{(d-1)}, \ldots, \hat{\boldsymbol{f}}_{(d-L)}\}$ are either fitted factors or forecast values, while d_0 is the last day observed in the sample. Note that although we use a single forecasting equation (5.10), the fitted factors and estimates of the loading matrices $\hat{\boldsymbol{\Lambda}}$ and the parameter matrices $\hat{\boldsymbol{B}}_l$ differ for the three estimation methods outlined. We label these three forecasting methods DFM1, DFM2 & DFM3, respectively.

In this section we introduced four different models to capture the serial dependence in the residual process $\boldsymbol{u}_{(d)}$. Three of them are factor based approaches. We now introduce the mean model that incorporates any external effects and that produces the residual process $\boldsymbol{u}_{(d)}$. We presented the residual models first to emphasize that the focus on our modelling exercise is on this part of our analysis.

5.2 Mean Component

We now specify the mean component $\mu_{h(t)}$ in equation (5.1). This varies over intraday periods, and we estimate the N regression means $\mu_1, \ldots, \mu_N$ separately. The covariate vector $\boldsymbol{x}_t$ can contain both calendar and meteorological effects. For example, the former typically includes a time trend, day type dummy variables and the day of the year $doy(t)$, while the latter can include air temperature variables, humidity and rainfall.

5 Application: Forecasting Energy Demand with Dynamic Factor Models

The covariates in our two studies include both continuous and dummy variables, which we denote as $\boldsymbol{c}_t = (c_{t1}, \ldots, c_{tp_c})^\top$ and $\boldsymbol{s}_t = (s_{t1}, \ldots, s_{tp_s})^\top$, respectively, so that $\boldsymbol{x}_t = (\boldsymbol{s}_t^\top, \boldsymbol{c}_t^\top)^\top$ with $p = p_s + p_c$ elements.

We model each of the dummy variables with a linear coefficient, but consider the impact of the continuous covariates c_{tj} as unknown smooth effects. The impact of some covariates on demand are known to be highly nonlinear and have been modeled previously as semiparametric effects; see Engle, Granger, Rice & Weiss (1986), Harvey & Koopman (1993) and Smith (2000). Here, the effect is denoted $f_j(c_{tj})$, with f_j a smooth but unknown function that is to be estimated. We use penalized splines (see Section 2.1), where each unknown function f_j is expressed as a linear combination of a large number of function basis terms, so that $f_j(c_{tj}) = \mathcal{B}_j(c_{tj})\boldsymbol{b}_j$. Here, $\mathcal{B}(c_{tj})$ is a cubic smoothing spline basis evaluated at c_{tj}, and $\boldsymbol{b}_j$ are the corresponding basis term coefficients. These are estimated using a penalized likelihood with quadratic penalties $\lambda_j \boldsymbol{b}_j^\top \boldsymbol{D}_j \boldsymbol{b}_j$ for each unknown function, where $\boldsymbol{D}_j$ is a fixed penalty matrix and λ_j is a smoothing parameter selected in a data driven way. This produces a function estimate that is simultaneously both flexible and smooth. The approach has a long history in the statistics literature (Wahba, 1990), and Ruppert, Wand & Carroll (2003, 2009) outline its implementation, practicability and the high quality of the resulting estimates.

During some periods of the day the impact of the calendar and meteorological variables on energy demand may not be additive. For example, Smith (2000) and Cottet & Smith (2003) demonstrate this using aggregate electricity demand and key meteorological variables from New South Wales, Australia. We therefore consider bivariate interaction effects in the model in the following three forms:

(i) All pairwise interactions between the dummy variables.

(ii) Interactions between the continuous variables and dummy variables in the form $s_{ti}\gamma_j(c_{tj})$, where γ_j is a smooth univariate function called a "varying coefficient" by Hastie & Tibshirani (1993).

(iii) Bivariate functions $f_{ij}(c_{ti}, c_{tj})$ for interactions between continuous covariates, using a penalized bivariate function basis.

5 Application: Forecasting Energy Demand with Dynamic Factor Models

To identify the model we assume that all smooth functions integrate to zero. Once a model is selected for each intraday period, estimation of the mean components is straightforward using the R-package mgcv provided by Wood (2006). Even though we consider only bivariate interactions, the resulting model can include a large number of smooth functions and linear coefficients. Therefore, we also identify a subset of effects from those above using the marginal BIC outlined below; see also Wager, Vaida & Kauermann (2007) or Greven & Kneib (2010). Calculation of this requires the assumption of a distribution for the disturbances u_t, and for this purpose we assume a first order Gaussian autoregression. Even though this is incorrect, it is likely to account for a sufficient amount of serial dependence so as to have little or no impact on the model choice. For example, Krivobokova & Kauermann (2007) demonstrate that the use of penalized spline smoothing for fitting smooth functions is robust with respect to mis-specifications of the correlation structure. When implementing this approach if an interaction effect is included, we maintain a hierarchy by including the univariate 'main effects' of both variables in the model. We demonstrate the ability of the method to select parsimonious, but flexible, mean components in our empirical work.

The components $\hat{\mu}_1, \ldots, \hat{\mu}_N$ are estimates of the marginal mean demand. These can be evaluated for future periods by employing future values of covariates. In the case of calendar effects they are known exactly, whereas for any meteorological variables forecasts need to be employed in practise. Demand is forecast over the horizon $t > T$ from the conditional mean

$$E(y_t|\mathcal{F}_T) = \hat{\mu}_{h(t)}(\boldsymbol{x}_t) + \hat{u}_t,$$

where $\hat{u}_t$ is obtained from the PAR model, or the dynamic factor model as in equation (5.10), and $\mathcal{F}_T$ is the filtration at time T.

The marginal Bayesian Information Criterion (BIC) that is used for the model selection routine is build on the fact that smooth functions can be fitted with penalized splines as originally suggested in Eilers & Marx (1996) and further developed in Ruppert, Wand & Carroll (2003). We illustrate the idea with the simple model

$$y_t = f_j(c_{tj}) + u_t,$$

with c_{tj} as continuous covariate and f_j as smooth but otherwise unspecified function. We fit the function by replacing f_j by some high dimensional basis $\mathcal{B}_j(x_{tj})\boldsymbol{b}_j$ with $\boldsymbol{b}_j$ as spline coefficient. A convenient and numerically stable basis is to use B-splines (de Boor, 1978) but in fact $\mathcal{B}_j$ can be any smoothing basis like cubic splines or thin plate splines, see Wood (2006). To achieve a smooth fit a quadratic penalty $\lambda_j \boldsymbol{b}_j^\top \boldsymbol{D}_j \boldsymbol{b}_j$ is imposed on the spline coefficients, where $\boldsymbol{D}_j$ is an appropriately chosen penalty matrix and λ_j the so called penalty parameter. One can now reformulate the penalty in a Bayesian view as prior distribution imposed on the spline coefficients. The quadratic form itself mirrors normality, i.e. $\boldsymbol{b}_j \sim \mathrm{N}(0, \sigma_j^2 \boldsymbol{D}_j^-)$ with $\boldsymbol{D}_j^-$ as (generalized) inverse and $\sigma_j^2 = \sigma_{th}^2/\lambda_j$. Assuming normality for residual u_t the resulting model is now of linear mixed model style

$$y_t | \boldsymbol{b}_j \sim \mathrm{N}(\mathcal{B}_j(c_{tj})\boldsymbol{b}_j, \sigma_u^2),$$
$$\boldsymbol{b}_j \sim \mathrm{N}(\boldsymbol{0}, \sigma_j^2 \boldsymbol{D}_j^-).$$

Integrating out the spline effects $\boldsymbol{b}_j$ gives the likelihood of the marginal model which can be maximized with respect to the remaining parameters, see Wand & Ormerod (2008) for details. In particular, the parameters can be estimated with maximum likelihood and the smoothing parameters result as a byproduct of the maximum likelihood estimation, see Wand (2003) for details and Kauermann, Krivobokova & Fahrmeir (2009) for asymptotic investigation of this relation, see also Kauermann (2005). Letting $\hat{l}$ denote the marginal log likelihood of the fitted model we define the marginal BIC with

$$\mathrm{BIC} = -2\hat{l} + p\log(T),$$

where p is the number of fitted parameters and T is the number of observations. The criterion can now be used for model selection in the conventional way.

Similar to the PAR model selection described in Section 5.1.1 we pursue a hierarchical stepwise model selection separately for each hour. Starting with the smallest model containing only an intercept and a smooth seasonal component, we check in each step if the inclusion or elimination of a variable or interaction effect leads to an improved BIC. We stop the model selection procedure if no improvement can be achieved.

5 Application: Forecasting Energy Demand with Dynamic Factor Models

In this section a selection strategy for the mean model was described. This is necessary to limit the numerical burden by only considering the relevant external effects. In the following two chapters the models presented up to now will be applied to both data examples mentioned above.

5.3 Wuppertal District Heat Demand

Our first application is to forecast district heating demand in the city of Wuppertal in North-Western Germany. Note that district heating demand in Wuppertal refers to both hot water supply and household heating. Heating is induced by steam (measured in tons per hour) taken from the district's steam network which is fed by two co-generating power stations. Steam has to be provided throughout the whole year and not only in winter months. In our example the majority of heating demand is from large scale consumers, with ten clients (out of 950 connected to the Wuppertal network) responsible for 38% of the total demand. We examine hourly demand between 1 January 2006 to 30 June 2009, where the first three years are used to estimate our models, and forecasts constructed for the last six months.

5.3.1 Mean Component

The covariates include the day of year $doy(t)$, current air temperature and its daily minimum and maximum, resulting in four continuous covariates. Dummy variables were introduced for five basic day types: Sundays, Mondays, Saturdays, Fridays and Tuesdays to Thursdays, where public holidays are classified as Sundays. A periodic cubic spline basis is used for the univariate function of $doy(t)$, and thin plate spline bases are used for the remaining univariate functions; see Section 2.1.4. For the bivariate interactions of the continuous variables a tensor product of the corresponding univariate bases is used, which is both a flexible bivariate basis and ensures the overall mean remains periodic in $doy(t)$. A parsimonious subset of these effects at each hour is selected by the

5 Application: Forecasting Energy Demand with Dynamic Factor Models

stepwise algorithm, in which the day type dummies are included or excluded as a block. The resulting optimal models are listed in Table 5.1, with effects that are not selected omitted.

The optimal models at each hour are similar to those in adjacent hours, which is a result of the strong diurnal structure of the determinants of heating demand and not an assumption of the model. For example, the day type is important between 05:00 and 20:00, but omitted overnight. Similarly, daily maximum temperature is important between 17:00 and 08:00 the next day, and usually as a bivariate interaction with $doy(t)$, whereas minimum temperature has less of an effect, and only between 11:00 and 21:00. Both current air temperature and $doy(t)$ are important throughout the day. Interestingly, the results suggest that the way in which meteorological conditions impact on heating demand not only differs over the time of day, but also over season.

Figure 5.1 plots some of the estimated effects. Tuesday to Thursday is considered as "baseline" demand, and panel (a) plots the estimated effect against the hour of the day h. Panel (b) plots deviations for the other four day type dummies over the period they are significant, and not surprisingly demand is lower during the weekends. Panels (c) and (d) plot the estimated univariate effects of $doy(t)$ and air temperature at three different times of the day. The former shows increased demand for heating as temperatures dip below around 15°C in a near linear fashion. This supports the popular "heating degree days" measure, which is a parametric function often used to account for the impact of air temperature on heating demand; for example, see Sailor & Muñoz (1997). Unsurprisingly, demand for heating is higher in winter than summer.

5.3.2 Time Series Components & Forecasts

Both the PAR and dynamic factor models were estimated using the residuals from the fitted mean component. Table 5.2 reports the optimal lag lengths $L_{1,h}$ and $L_{2,h}$ for the PAR. The maximum value considered for both lags was 10, although selected intraday lags are three hours or less, and inter-day lags are two days or less. In all three factor models we found that three factors explained in excess of 90% of the total variation, providing a substantial dimension reduction for the $N = 24$ dimensional multivariate

5 Application: Forecasting Energy Demand with Dynamic Factor Models

Table 5.1: Effects in the marginal mean model for Wuppertal heat demand, as selected by BIC. Each row corresponds to a different hour, and inclusion of a component is denoted with "•" and exclusion by "−"; effects not selected at any hour are omitted from the table. An adjusted R^2 value for each regression equation is also given.

Hour	Main Effects					Interactions		Adj. R^2
	doy	Max. Temp.	Min. Temp.	Temp.	Day Type	Max.Temp. & doy	Temp. & doy	
00:00	•	•	−	•	−	−	•	0.906
01:00	•	•	−	•	−	•	−	0.906
02:00	•	•	−	•	−	•	−	0.909
03:00	•	•	−	•	−	•	•	0.910
04:00	•	•	−	•	−	•	−	0.909
05:00	•	•	−	•	•	•	−	0.908
06:00	•	•	−	•	•	•	−	0.905
07:00	•	•	−	•	•	•	−	0.902
08:00	•	•	−	•	•	•	−	0.904
09:00	•	−	−	•	•	−	•	0.908
10:00	•	−	−	•	•	−	−	0.907
11:00	•	−	•	•	•	−	•	0.914
12:00	•	−	•	•	•	−	•	0.917
13:00	•	−	•	•	•	−	•	0.912
14:00	•	−	•	•	•	−	•	0.921
15:00	•	−	•	•	•	−	•	0.925
16:00	•	−	•	•	•	−	•	0.924
17:00	•	•	•	•	•	−	•	0.924
18:00	•	•	−	•	•	−	•	0.926
19:00	•	•	•	•	•	−	•	0.929
20:00	•	•	−	•	•	•	−	0.924
21:00	•	•	•	•	−	•	−	0.922
22:00	•	•	−	•	−	•	−	0.922
23:00	•	•	−	•	−	•	−	0.918

5 Application: Forecasting Energy Demand with Dynamic Factor Models

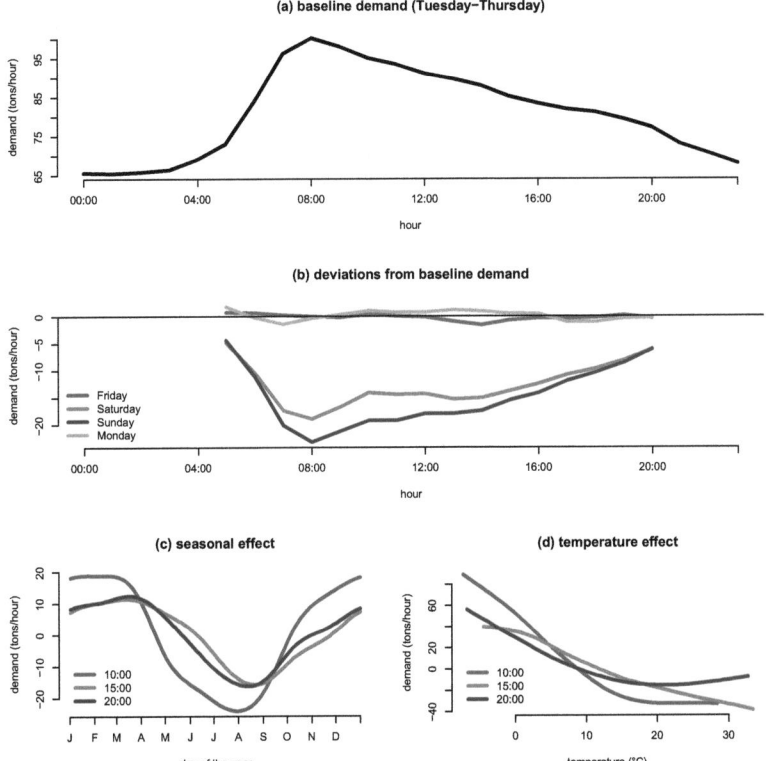

Figure 5.1: Estimated components from the mean model for heating demand in Wuppertal. Panel (a): Estimated baseline demand on Tuesday–Thursday against hour of the day h. Panel (b): Deviations from the baseline demand for other day types. Panels (c) and (d): Smooth seasonal and temperature effects at three times of the day (10:00, 15:00 and 20:00).

5 Application: Forecasting Energy Demand with Dynamic Factor Models

Hour h	00:00	01:00	02:00	03:00	04:00	05:00	06:00	07:00	08:00	09:00	10:00	11:00
Intraday Lag $L_{h,1}$	1	3	1	1	2	3	1	1	3	1	1	3
Inter-day Lag $L_{h,2}$	1	0	0	1	1	1	2	1	2	2	1	1
Hour h	12:00	13:00	14:00	15:00	16:00	17:00	18:00	19:00	20:00	21:00	22:00	23:00
Intraday Lag $L_{h,1}$	1	2	3	1	2	3	3	2	1	2	2	1
Inter-day Lag $L_{h,2}$	0	0	1	1	2	1	1	1	1	1	1	1

Table 5.2: Lags chosen for the PAR model applied to the heat demand data.

time series.

Forecasts are made between one and seven days ahead at the hourly resolution over the first six months of 2009, and the absolute percentage errors (APEs) computed. Table 5.3 reports the mean, median and 90th percentiles of the daily mean APEs (that is the daily MAPEs) for five models. Because higher daily MAPEs correspond to less accurate forecasts, the 90th percentile corresponds to typical poor forecast days. Forecasts were first produced using the marginal means $\{\hat{\mu}_{h(t)}(\boldsymbol{x}_t); t > T\}$ as a benchmark, which are the same over the forecast horizon. For comparison, marginal mean forecasts were also re-computed without the meteorological variables, and are also quoted in Table 5.3. Unsurprisingly, the forecasts that take into account meteorological conditions are an improvement over those that do not, and underline the potential value of incorporating weather forecasts into energy forecasting models for such horizons- a common practise in energy utilities. The optimal lag length of the vector autoregression of the factors in equation (5.6) is $L = 1$ day.

The marginal mean (with weather) is taken as a reference, and the summaries for the time series model forecasts quoted as percentage reductions against this benchmark. All time series models provide a substantial and similar improvement in forecast accuracy one day ahead, reducing the average daily MAPEs between 51% and 54%. For longer forecast horizons the time series models continue to provide substantial improvements, but the dynamic factor models dominate the PAR. In particular, DFM2 is the most reliable when considering the 90th percentiles and the best performing overall. On average, forecasts seven days ahead can be improved by 34.4% by employing DFM2, even when already accounting for the nonlinear effect of key meteorological variables and periodicity.

5 Application: Forecasting Energy Demand with Dynamic Factor Models

Method	+1	+2	+3	+4	+5	+6	+7
Mean of Daily MAPE Values							
Marginal Mean			0.173 (0.183 without weather)				
PAR	-51.1%	-45.5%	-41.2%	-36.9%	-32.3%	-27.8%	-24.4%
DFM1	-53.4%	-47.6%	-45.4%	-43.4%	-39.4%	-35.3%	-32.1%
DFM2	**-53.9%**	-48.6%	-46.8%	**-45.3%**	**-41.6%**	**-37.6%**	**-34.4%**
DFM3	-53.3%	**-49.0%**	**-46.9%**	-44.5%	-40.1%	-35.7%	-32.3%
Median of Daily MAPE Values							
Marginal Mean			0.156 (0.164 without weather)				
PAR	-50.21%	-44.36%	-40.64%	-34.45%	-31.04%	-27.93%	-24.06%
DFM1	-52.08%	-48.34%	-47.01%	-45.03%	-40.60%	-37.34%	-35.57%
DFM2	**-53.88%**	-49.41%	-47.36%	-45.79%	-42.89%	**-39.26%**	**-38.21%**
DFM3	-53.76%	**-51.60%**	**-49.18%**	**-48.41%**	**-44.73%**	-38.09%	-35.23%
90th Percentile of Daily MAPE Values							
Marginal Mean			0.297 (0.336 without weather)				
PAR	-56.31%	-49.30%	-41.07%	-36.14%	-31.13%	-24.51%	-22.31%
DFM1	-56.21%	-49.55%	-47.05%	-40.99%	-38.71%	-34.89%	-26.82%
DFM2	**-56.90%**	**-52.07%**	**-51.35%**	**-44.22%**	**-41.00%**	**-37.64%**	**-29.26%**
DFM3	-56.78%	-49.88%	-44.83%	-39.99%	-37.66%	-33.92%	-27.52%

Table 5.3: Summaries of the daily MAPEs of hourly forecasts for heating demand in Wuppertal using five different methods. The summaries are the mean, median and 90th percentiles, and the results are given for forecast horizons of between one and seven days ahead. The MAPE is reported when using the marginal mean as a forecast and is the same regardless of forecast horizon. Results for the four time series models differ by forecast horizon and are quoted as percentage reductions relative to the mean model summaries. The result of the best performing method is in bold.

5 Application: Forecasting Energy Demand with Dynamic Factor Models

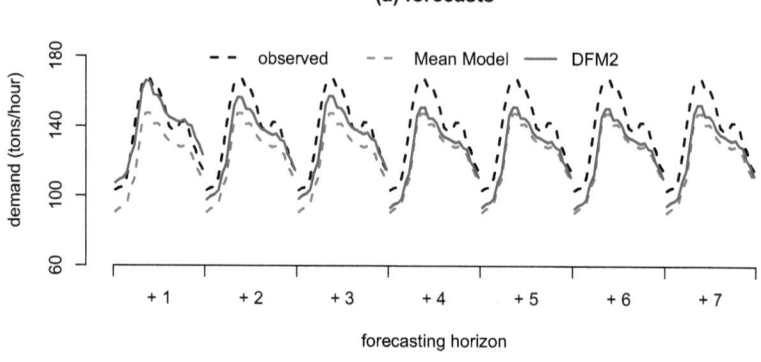

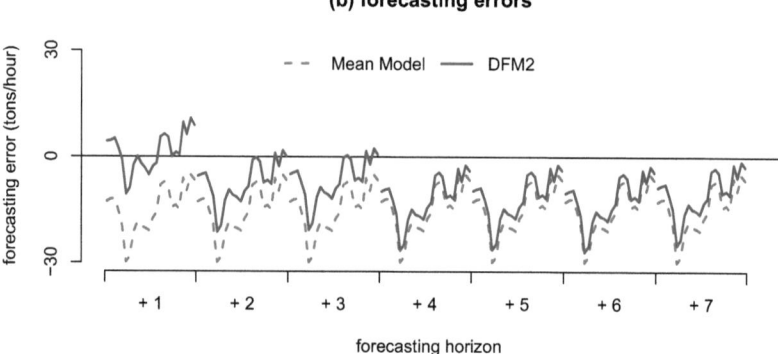

Figure 5.2: Panel (a): forecasts of heating demand in Wuppertal for Tuesday 20 January 2009 made between one to seven days ahead using both the marginal mean and DFM2 dynamic factor model. Also included is the observed demand. Panel (b): forecast errors corresponding to the two forecasts.

5 Application: Forecasting Energy Demand with Dynamic Factor Models

As an illustration, Figure 5.2 gives the hourly forecasts for Tuesday, 20 January 2009, made one to seven days previously. Forecasts using the marginal mean and the best performing time series model (DFM2) are plotted in panel (a), along with observed heating demand. The corresponding forecast errors are also plotted in panel (b), and the time series forecasts can be seen to converge to the marginal mean, which is because the time series is stationary.

5.4 Victorian Electricity Demand

In the absence of demand shedding, electricity demand is equal to the load on a power system. Forecasting electricity demand at an intraday resolution over a variety of horizons is an important problem faced by electricity utilities worldwide. In this section we forecast aggregate electricity demand in the Australian state of Victoria at a half-hourly resolution. Victoria covers an area of 227,420 km^2 and weather conditions can vary substantially across the state. However, because approximately 74% of the population live in the capital city of Melbourne and its immediate surroundings, we use meteorological conditions at this location. We employ half-hourly observations from the meteorological stations Viewbank, Essendon and Moorabin, which are the three stations located in central Melbourne. Because of their close proximity, there are only very minor variations between the three sets of observations. The series feature short tracts of missing data, and to construct a complete series we use the average of observations across the three monitoring stations. We consider demand between 1 January 2006 to 30 September 2009, and employ the first three years of data as a training sample, from which we make forecasts for the last nine months.

5.4.1 Mean Component

The meteorological variables used are air temperature, relative humidity, precipitation and windspeed. We also include the daily maximum and minimum air temperature and $doy(t)$, resulting in seven continuous covariates. The same bases outlined in Section 5.3.1

Table 5.4: Effects in the marginal mean model for Victorian electricity demand, as selected by BIC. Each row corresponds to a different hour, and inclusion of a component is denoted with "•" and exclusion by "—"; effects not selected at any hour are omitted from the table. An adjusted R^2 value for each regression equation is also given.

Hour	Main Effects					Interactions				Adj. R^2
	doy	Max.Temp	Min.Temp	Temp.	Day Type	Max.Temp & doy	Temp. & doy	Max.Temp & Day Type	Temp. & Day Type	
00:00	•	•	—	•	•	•	•	—	—	0.879
00:30	•	•	—	•	•	•	•	—	—	0.900
01:00	•	•	—	•	•	•	•	—	—	0.907
01:30	•	•	—	•	•	•	•	—	—	0.905
02:00	•	•	—	•	•	•	•	—	—	0.914
02:30	•	•	—	•	•	•	•	—	—	0.923
03:00	•	•	—	•	•	•	•	—	—	0.921
03:30	•	•	—	•	•	•	•	—	—	0.909
04:00	•	•	—	•	•	•	•	—	—	0.889
04:30	•	•	—	•	•	•	•	—	—	0.857
05:00	•	•	—	•	•	•	•	—	—	0.827
05:30	•	•	—	•	•	•	•	—	—	0.801
06:00	•	•	—	•	•	•	•	—	—	0.807
06:30	•	•	—	•	•	•	•	—	—	0.850
07:00	•	•	—	•	•	•	•	—	—	0.889
07:30	•	•	—	•	•	•	•	—	—	0.920
08:00	•	•	—	•	•	•	•	•	•	0.937
08:30	•	•	—	•	•	•	•	•	•	0.940
09:00	•	—	—	•	•	—	•	—	—	0.940
09:30	•	—	—	•	•	—	•	—	—	0.929
10:00	•	•	—	•	•	•	•	—	—	0.925
10:30	•	•	—	•	•	•	•	—	—	0.919
11:00	•	•	—	•	•	•	•	—	—	0.918
11:30	•	•	—	•	•	•	•	—	—	0.916
12:00	•	•	—	•	•	•	—			0.916
12:30	•	•	—	•	•	•	—			0.918
13:00	•	•	—	•	•	•	—			0.921
13:30	•	•	—	•	•	•	—			0.920
14:00	•	•	—	•	•	—	•			0.923
14:30	•	•	—	•	•	—	•			0.925
15:00	•	•	—	•	•	—	•			0.926
15:30	•	•	—	•	•	•	•			0.917
16:00	•	•	—	•	•	•	•			0.913
16:30	•	•	—	•	•	•	•			0.916
17:00	•	•	—	•	•	•	•			0.913
17:30	•	•	—	•	•	•	•			0.914
18:00	•	•	—	•	•	•	•			0.928
18:30	•	•	—	•	•	•	•			0.933
19:00	•	•	—	•	•	—	•			0.932
19:30	•	•	—	•	•	—	•			0.931
20:00	•	•	—	•	•	—	•			0.924
20:30	•	•	—	•	•	—	•			0.914
21:00	•	•	—	•	•	—	•			0.911
21:30	•	•	—	•	•	—	•			0.910
22:00	•	•	—	•	•	•	•			0.912
22:30	•	•	—	•	•	•	•			0.913
23:00	•	•	—	•	•	•	•			0.912
23:30	•	•	—	•	•	•	•			0.915

5 Application: Forecasting Energy Demand with Dynamic Factor Models

were used, along with the same day type dummies. Table 5.4 shows the optimal model identified by the stepwise algorithm, which omits precipitation, humidity and windspeed at all times of the day. In comparison, Panagiotelis & Smith (2008b) found these meteorological variables to have a minor effect on Victorian electricity demand using a similar semiparametric regression model, and the reason for the difference is two-fold. First, daily maximum and minimum air temperature were not considered by Panagiotelis & Smith (2008b), and if these are omitted here then humidity, the second strongest meteorological variable identified by Panagiotelis & Smith (2008b), also features in our optimal model. However, forecasts of both daily maximum and minimum air temperature are readily available, so it is preferable to employ these. Second, the BIC is an approximation to the Bayesian posterior model probability, whereas the method of Panagiotelis & Smith (2008b) computes these exactly given their choice of Bayesian priors. While BIC is well-known to select sometimes overly parsimonious models compared to the exact posterior model probability, this is less problematic in our approach because the time series component can capture any residual weather effects.

The importance of seasonality, day type and air temperature in determining electricity demand is well-documented for many locations (Pardo, Meneu & Valor, 2002; Moral-Carcedo & Vicens-Otero, 2005), and our empirical results confirm these findings. Figure 5.3 plots the same effects as those presented for the German heating demand study. The deviations in the day type effects in panel (b), from the baseline case in panel (a), are now prevalent at all hours of the day. The relationship with air temperature in panel (d) now displays the combined impact of both heating and cooling demand, with an increase in electricity demand for high temperatures as well as low; air-conditioning is commonplace in Victoria. The point of minimum demand varies at different times of the day, as also documented for other electricity demand series by a number of authors; for example, see Cottet & Smith (2003) and Moral-Carcedo & Vicens-Otero (2005). The seasonal component in Figure 5.3(c) is markedly different from that in Figure 5.1(c) for the same reason.

5 Application: Forecasting Energy Demand with Dynamic Factor Models

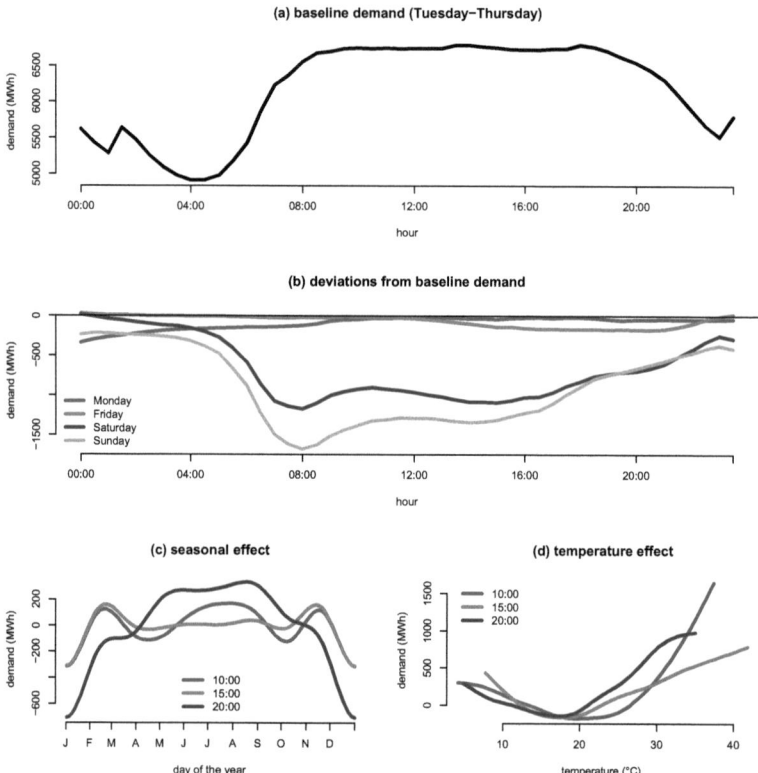

Figure 5.3: Estimated components from the mean model for Victorian electricity demand. Panel (a): Estimated baseline demand on Tuesday–Thursday against hour of the day h. Panel (b): Deviations from the baseline demand for other day types. Panels (c) and (d): Smooth seasonal and temperature effects at three times of the day (10:00, 15:00 and 20:00).

5 Application: Forecasting Energy Demand with Dynamic Factor Models

Table 5.5: Lags chosen for the PAR model applied to the electricity demand data.

Hour h	00:00	00:30	01:00	01:30	02:00	02:30	03:00	03:30	04:00	04:30	05:00	05:30
Intraday Lag $L_{h,1}$	2	2	1	5	3	2	2	1	1	1	5	4
Inter-day Lag $L_{h,2}$	1	–	1	1	1	–	1	3	3	1	1	1
Hour h	06:00	06:30	07:00	07:30	08:00	08:30	09:00	09:30	10:00	10:30	11:00	11:30
Intraday Lag $L_{h,1}$	4	1	2	2	2	1	2	1	2	2	3	1
Inter-day Lag $L_{h,2}$	1	1	3	1	1	1	1	1	1	–	–	1
Hour h	12:00	12:30	13:00	13:30	14:00	14:30	15:00	15:30	16:00	16:30	17:00	17:30
Intraday Lag $L_{h,1}$	2	1	1	2	2	2	2	1	1	4	4	1
Inter-day Lag $L_{h,2}$	1	1	1	–	–	–	1	–	–	–	–	1
Hour h	18:00	18:30	19:00	19:30	20:00	20:30	21:00	21:30	22:00	22:30	23:00	23:30
Intraday Lag $L_{h,1}$	4	2	3	2	1	1	2	1	3	1	2	3
Inter-day Lag $L_{h,2}$	1	3	2	1	1	1	1	1	3	1	1	2

5.4.2 Time Series Component & Forecasts

The intraday lag length $L_{2,h}$ can be as long as 5 periods (2.5 hours), while inter-day lag length $L_{2,h}$ can extend to 3 days at some hours of the day h (see Table 5.5). The latter is probably due to the combined effect of thermal inertia and any residual weather effect not captured by the mean component. In all three dynamic factor models $M = 4$ factors are found to explain in excess of 90% of the total variance. The optimal lag length of the factor vector autoregression in equation (5.6) is $L = 3$, which matches $\max_h\{L_{2,h}\} = 3$ in the PAR model.

Using all models we compute half-hourly forecasts of electricity demand up to seven days ahead. Table 5.6 provides the summaries of daily MAPE values in the same format as the German heating demand example. Incorporating weather variables into the mean component reduces the mean daily MAPEs substantially from 4.77% to 2.91%, and the 90th percentile (typical poor forecast) of daily MAPEs from 9.1% to 5.2%. This is a greater improvement than that observed for Wuppertal heating demand, which is likely due to two reasons. First, district heating in Wuppertal is dominated by a small number of large users who appear less weather sensitive than the approximately two million households that dominate aggregate electricity demand in Victoria. Second, electricity demand also comprises cooling demand, and this is both extensive and highly variable Victoria, which experiences frequent extreme temperature variations during summer.

5 Application: Forecasting Energy Demand with Dynamic Factor Models

Method	+1	+2	+3	+4	+5	+6	+7
			Days Ahead				
Mean of Daily MAPE Values							
Marginal Mean			0.0291 (0.0477 without weather)				
PAR	-25.80%	-19.39%	-16.04%	-13.35%	-10.47%	-8.09%	-6.06%
DFM1	**-28.74%**	-19.44%	-15.07%	-11.48%	-9.45%	-7.81%	-6.44%
DFM2	-27.07%	-20.62%	-17.02%	-13.56%	-11.43%	-9.96%	**-8.10%**
DFM3	-27.79%	**-21.25%**	**-17.63%**	**-13.88%**	**-11.57%**	**-10.07%**	**-8.10%**
Median of Daily MAPE Values							
Marginal Mean			0.0244 (0.0379 without weather)				
PAR	-21.81%	-11.15%	-9.46%	-6.94%	-6.65%	-6.13%	-4.71%
DFM1	**-25.85%**	-16.86%	-13.82%	-8.21%	-5.30%	-6.73%	-7.73%
DFM2	-25.10%	-20.43%	-16.54%	-10.53%	-7.11%	-7.76%	-7.38%
DFM3	-24.41%	**-21.43%**	**-17.30%**	**-11.16%**	**-7.92%**	**-8.25%**	**-8.93%**
90th Percentile of Daily MAPE Values							
Marginal Mean			0.0520 (0.0910 without weather)				
PAR	-34.92%	**-28.52%**	**-22.85%**	-17.52%	-12.71%	-8.85%	-5.56%
DFM1	-35.84%	-28.04%	-20.41%	**-18.96%**	-15.80%	-9.51%	-7.30%
DFM2	**-36.47%**	-28.36%	-20.00%	-18.44%	**-15.23%**	**-10.11%**	**-7.78%**
DFM3	-35.46%	-26.74%	-17.78%	-16.35%	-10.95%	-8.11%	-7.31%

Table 5.6: Summaries of the daily MAPEs of half-hourly forecasts for electricity demand in Victoria using five different methods. The summaries are the mean, median and 90th percentiles, and the results are given for forecast horizons of between one and seven days ahead. The MAPE is reported when using the marginal mean as a forecast and is the same regardless of forecast horizon. Results for the four time series models differ by forecast horizon and are quoted as percentage reductions relative to the mean model summaries. The result of the best performing method is in bold.

5 Application: Forecasting Energy Demand with Dynamic Factor Models

	Days Ahead						
Method	+1	+2	+3	+4	+5	+6	+7
PAR	$30,602	$23,520	$19,948	$17,005	$13,667	$10,453	$7,855
DFM1	$38,473	$31,153	$23,300	$16,517	$13,735	$11,272	$8,410
DFM2	$39,142	$32,152	$24,077	$16,645	$14,086	$11,338	$8,183
DFM3	$39,976	$33,093	$26,136	$17,635	$14,762	$12,012	$9,040

Table 5.7: Average daily monetary value of the time series model forecasts for electricity demand in Victoria over-and-above the marginal mean model. Results are given for forecast horizons of between one and seven days ahead.

All time series models provide a substantial improvement in forecast accuracy over all horizons. While there is less difference between the PAR and dynamic factor models than in the previous application, again the dynamic factor models dominate, except for the 90th percentile of daily MAPEs at horizons of 2 and 3 days ahead. On average, forecasts from the DFM3 model dominate for horizons of between 2 and 7 days ahead, although the DFM2 model is more accurate when considering the 90th percentile.

A monetary value of the improved forecast accuracy provided by the time series models can be computed using the wholesale spot price as follows. Let $\hat{y}_t$ be the forecast demand (in MW per hour) from using a time series component, and P_t be the half-hourly spot price (in $ per MW per hour). Then the monetary benefit of the improved forecast at half hour t is

$$B_t = P_t(|y_t - \mu_{h(t)}(\boldsymbol{x}_t)| - |y_t - \hat{y}_t|)/2,$$

and the daily benefit is the summation of the half-hourly benefits during a day. Table 5.7 provides the average daily monetary values computed over the nine month forecast horizon, showing that this improvement has a sizable expected monetary value. Figure 5.4.2 illustrates the ability of the methodology to provide forecasts over a one week horizon. Panel (a) plots forecasts of half-hourly demand for Sunday, 5 July 2009, made from one to seven days previously using both the DFM3 model and the marginal mean.

5 Application: Forecasting Energy Demand with Dynamic Factor Models

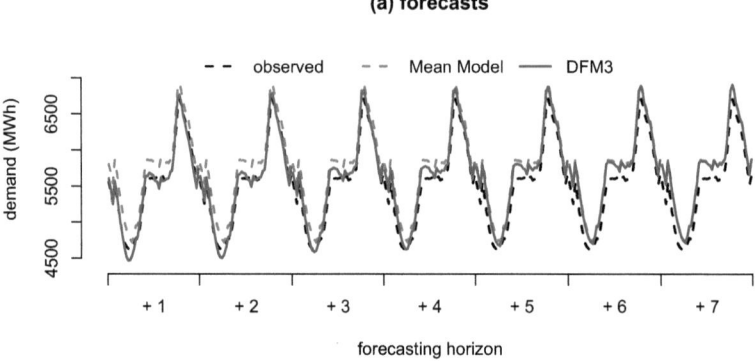

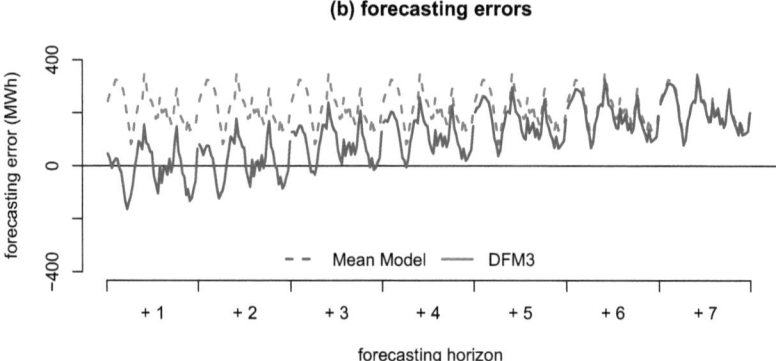

Figure 5.4: Panel (a): Forecasts of Victorian electricity demand for Sunday, 5 July 2009, made between one to seven days ahead using both the marginal mean and DFM3 dynamic factor model. Also included is the observed demand. Panel (b): forecast errors corresponding to the two forecasts.

5 Application: Forecasting Energy Demand with Dynamic Factor Models

Panel (b) provides the corresponding errors and demonstrates that the benefits of using the dynamic factor model extend over the entire horizon.

5.5 Results

The approach we outline in this chapter can be applied to forecasting intraday demand for electricity, gas and steam. The penalized spline smoothing methodology is well suited to estimating complex semiparametric regression models of the type considered here. It is straightforward to implement using the public domain software R, along with the maximum likelihood estimators for the multivariate time series models. In our empirical work we employ meteorological variables and show they can greatly enhance the accuracy of demand forecasts over a one week horizon. Of course, in practise forecast values for meteorological variables would have to be used, which is an approach widely employed by electricity utilities, although the weather component could be omitted if required. In the empirical work we find the expected relationship between current air temperature and both heating and electricity demand. We confirm previous studies (Pardo, Meneu & Valor, 2002; Cottet & Smith, 2003) that show that form of the relationship varies during both the day and season, so that any heating and cooling degree day measures constructed should also vary by day and season. Interestingly, maximum and minimum daily temperature are also important, although they have often been overlooked in previous studies.

After accounting for calendar and weather effects in the marginal mean models for each intraday period, there is still substantial residual serial dependence. Modeling this using the multivariate time series models provides a consistent improvement in forecast accuracy, as well as making the whole approach robust to either mis-specification of the mean, or omission of weather components. Last, one result of our empirical work that was a surprise to us was that the dynamic factor models consistently out-perform the PAR.

6 Discussion and Perspective

Within the framework of dynamic water temperature management we argued that it is helpful to know if the current season is running ahead or behind an average year. Given this information conclusions can be drawn how much waste heat can be dissipated into the stream water without interfering with the river wild life. We employed landmark-based curve registration first suggested by Kneip & Gasser (1992) which is widely recognized as the most powerful curve registration model in the literature. However, as we treat time series data instead of complete functional observations and as we wanted to develop a procedure that can be applied online, the methodology of Kneip & Gasser had to be modified appropriately. We defined four different online landmark criteria that partly exploited the multivariate structure of the data at hand. From the resulting landmarks the corresponding reference points in the average year can be estimated and a time-warping function can be derived which was done by monotone smoothing with quadratic programming. However, our approach did not fully match the ecological data that we used to validate our results. This may partly be due to the small amount of spawning data that we were able to obtain. More observations should be collected in the future to get a better idea of the performance of our approach. Furthermore, our bootstrap routine demonstrated that the landmarks coming from the four criteria differ in variability. As we used smoothing techniques to calculate the time-warping functions we did not include weights for the different landmarks. However, this aspect may be worth further examination. Overall, the procedure seems to work fine for the presented dataset but it is not a methodology that can be applied to any online monitored seasonal time series. It remains a specialized approach that might be extended to other river water temperature data.

The applications of approximate dynamic factor models presented in Chapters 4 and

6 Discussion and Perspective

5 turned out to be quite successful.

In the case of water temperature forecast we showed that traditional univariate forecasting models are easily outperformed by our approaches. This is not too surprising as our models are apt to handle a huge amount of information, far more than could be processed using traditional time series models. However, this additional information can be obtained at low cost and the resulting improvements strongly emphasize the superiority of our dynamic factor model. The modelling routine was split into two parts. First we extracted the houly annual mean temperature course to guarantee first-order stationarity. This was done for both, water and air temperature data. We then formulated a combined model that was able to predict water temperature factor scores based on water and air temperature factors of previous days (and in the case of air temperature the same day, as well). We discussed three types of factor estimation and it turned out that the straight forwart least squares approach yielded the best result closely followed by the competing maximum likelihood based routines. The classical autoregressive model that we used as benchmark was outperformed, as well. It might be worthwhile to test the presented models on other stream water datasets to assess if the factor models prevail.

For the energy demand forecast we found similar results. For both, heat and electricity demand, the dynamic factor models dominated the competing periodic autoregressive model. The improvements were not so clear as in the previous chapter where water temperature was treated. But this is not a surprise as at least high-resolution electricity forecasts are already broadly discussed in the literature. We, again, pursued a two-stage modelling where this time we incorporated the external effects in the mean model rather than on the factor level. The examined models where built on the residual process and we did an out-of-sample forecast to evaluate the performances. Although all three factor models beat the benchmark it is not so easy to say which one of them is the best as they perform more or less equally well. This may shed some positive light on the first least squares based method that closely follows the suggestions of Stock & Watson (2002a, 2002b) and in our application performed only marginally worse than the other approaches but is far easier to implement. Applications to further data would be necessary to identify the overall best performing candidate.

References

Bigot, J. (2006). Landmark-based registration of curves via the continous wavelet transform. *Journal of Computational and Graphical Statistics* **15**, 542–564.

de Boor, C. (1978). *A Practical Guide to Splines.* Berlin: Springer.

Borak, S. and Weron, R. (2008). A semiparametric factor model for electricity forward curve dynamics. SFB 649 Discussion Paper 2008-050.

Box, G. E. P., Jenkins, G. M., and Reinsel, G. C. (1994). *Time series analysis* (3rd ed.). Englewood Cliffs, NJ: Prentice Hall.

Breitung, J. and Eickmeier, S. (2006). Dynamic factor models. *AStA Advances in Statistical Analysis* **90**, 27–42.

Breslow, N. E. and Clayton, D. G. (1993). Approximate inference in generalized linear mixed models. *Journal of the American Statistical Association* **88**(421), 9–25.

Brillinger, D. R. (1981). *Time Series: Data Analysis and Theory.* Holden-Day.

Broszkiewicz-Suwaja, E., Makagon, A., Weron, R., and Wylomanska, A. (2004). On detecting and modeling periodic correlation in financial data. *Physica A* **336**, 196–205.

Byeong, U. P., Mammen, E., Härdle, W., and Borak, S. (2009). Time series modelling with semiparametric factor dynamics. *Journal of the American Statistical Association* **104**, 284–298.

Caissie, D. (2006). The thermal regime of rivers: a review. *Freshwater Biology* **51**, 1389–1406.

REFERENCES

Caissie, D., El-Jabi, N., and Satish, M. G. (2001). Modelling of maximum daily water temperatures in a small stream using air temperatures. *Journal of Hydrology* **251**, 14–28.

Caissie, D., El-Jabi, N., and St-Hilaire, A. (1998). Stochastic modelling of water temperatures in a small stream using air to water relations. *Canadian Journal of Civilian Engineering* **25**, 250–260.

Caissie, D., Satish, M. G., and El-Jabi, N. (2005). Predicting river water temperatures using the equilibrium temperature concept with application on Miramichi River catchments (New Brunswick, Canada). *Hydrological Processes* **19**, 2137–2159.

Cancelo, J. R., Espasa, A., and Grafe, R. (2008). Forecasting the electricity load from one day to one week ahead for the spanish system operator. *International Journal of Forecasting* **24**, 588–602.

Cardot, H., Faivre, R., and Goulard, M. (2003). Functional approaches for predicting land use with the temporal evolution of coarse resolution remote sensing data. *Journal of Applied Statistics* **30**, 1185–1199.

Chamberlain, G. (1983). Funds, factors, and diversification in arbitrage pricing models. *Econometrica* **51**, 1281–1304.

Chamberlain, G. and Rothschild, M. (1983). Arbitrage factor structure, and mean-variance analysis of large asset markets. *Econometrica* **51**, 1281–1304.

Chiou, J.-M., Müller, H.-G., and Wang, J.-L. (2003). Functional quasi-likelihood regression models with smooth random effects. *Journal of the Royal Statistical Society – Series B* **65**(2), 405–423.

Cluis, D. (1972). Relationship between stream water temperature and ambient air temperature — a simple autoregressive model for mean daily stream water temperature fluctuations. *Nordic Hydrology* **3**, 65–71.

Connor, G. and Korajczyk, R. A. (1986). Performance measurement with the arbitrage pricing theory. *Journal of Financial Economics* **15**, 373–394.

REFERENCES

Connor, G. and Korajczyk, R. A. (1993). A test for the number of factors in an approximate factor model. *Journal of Finance* **XLVIII**, 1263–1291.

Cottet, R. and Smith, M. S. (2003). Bayesian modeling and forecasting of intraday electricity load. *Journal of the American Statistical Association* **98**, 839–849.

Darbellay, G. and Slama, M. (2000). Forecasting the short-term demand for electricity: Do neural networks stand a better chance? *International Journal of Forecasting* **16**, 71–83.

Dempster, A. P., Laird, N. M., and Rubin, D. B. (1977). Maximum likelihood from incomplete data via the EM algorithm. *Journal of the Royal Statistical Society - Series B* **39**, 1–38.

Dordonnat, V., Koopman, S. J., Ooms, M., Dessertaine, A., and Collet, J. (2008). An hourly periodic state space model for modelling french national electricity load. *International Journal of Forecasting* **24**, 566–587.

Dotzauer, E. (2002). Simple model for prediction of loads in district-heating systems. *Applied Energy* **73**, 277–284.

Duchon, J. (1977). Splines minimizing rotation-invariant semi-norms in Sobolev spaces. In W. Schemp & K. Zeller (Eds.), *Construction Theory of Functions of Several Variables*, pp. 85–100. Berlin: Springer.

Eilers, P. H. C. and Marx, B. D. (1996). Flexible smoothing with B-splines and penalties. *Statistical Science* **11**(2), 89–121.

Engle, R., Granger, W., Rice, J., and Weiss, A. (1986). Semiparametric estimates of the relationship between weather and electricity sales. *Journal of the American Statistical Association* **81**, 310–320.

Erbas, B., Hyndman, R. J., and Gertig, D. M. (2007). Forecasting age-specific breast cancer mortality using functional data models. *Statistics in Medicine* **26**, 458–470.

Ferraty, F. and Vieu, P. (2006). *Nonparametric Functional Data Analysis: Theory and Practice*. New York: Springer.

REFERENCES

Forni, M., Hallin, M., Lippi, M., and Reichlin, L. (2000). The generalized dynamic factor model: identification and estimation. *The Review of Economics and Statistics* **82**, 540–554.

Forni, M., Hallin, M., Lippi, M., and Reichlin, L. (2004). The generalized dynamic factor model: consistency and rates. *Journal of Econometrics* **119**, 231–255.

Forni, M., Hallin, M., Lippi, M., and Reichlin, L. (2005). The generalized dynamic factor model: One-sided estimation and forecasting. *Journal of the American Statistical Association* **471**, 830–840.

Forni, M. and Reichlin, L. (1998). Dynamic common factors in large cross-sections. *Empirical Economics* **65**, 453–473.

Franses, P. H. and Paap, R. (2004). *Periodic Time Series Models*. Advanced Texts in Econometrics. Oxford University Press.

Garnett, J. C. (1919). General ability, cleverness, and purpose. *British Journal of Psychology* **9**, 345–366.

Gasser, T. and Kneip, A. (1995). Searching for structure in curve samples. *Journal of the American Statistical Association* **90**, 1179–1188.

Gasser, T., Kneip, A., Binding, A., Prader, A., and Molinari, L. (1991a). The dynamics of linear growth in distance, velocity and acceleration. *Annals of Human Biology* **18**, 187–205.

Gasser, T., Kneip, A., Ziegler, P., Largo, R., Molinari, L., and Prader, A. (1991b). The dynamics of growth of height in distance, velocity and acceleration. *Annals of Human Biology* **18**, 449–461.

Gasser, T., Kneip, A., Ziegler, P., Largo, R., and Prader, A. (1990). A method for determining the dynamics and intensity of average growth. *Annals of Human Biology* **17**, 459–474.

Gervini, G. and Gasser, T. (2004). Self-modelling warping functions. *Journal of the Royal Statistical Society - Series B* **66**, 959–971.

REFERENCES

Gervini, G. and Gasser, T. (2005). Nonparametric maximum likelihood estimation of the structural mean of a sample of curves. *Biometrika* 92(4), 801–820.

Geweke, J. (1977). The dynamic factor analysis of economic time series. In D. J. Aigner & A. S. Goldberger (Eds.), *Latent Variables in Socio-Economic Models*, Chapter 19. Amsterdam: North-Holland.

Green, P. J. (1987). Penalized likelihood for general semi-parametric regression models. *International Statistical Review* 55, 245–259.

Green, P. J. and Silverman, B. W. (1994). *Nonparametric Regression and Generalized Linear Models*. Chapman & Hall.

Greven, S. and Kneib, T. (2010). On the behavior of marginal and conditional AIC in linear mixed models. *Biometrika* 97, 773–789.

Guthrie, G. and Videbeck, S. (2007). Electricity spot price dynamics: Beyond financial models. *Energy Policy* 35, 5614–5621.

Hallin, M. and Liška, R. (2007). Determining the number of factors in the general dynamic factor model. *Journal of the American Statistical Association* 102, 603–617.

Härdle, W., Horowitz, J. L., and Kreiss, J.-P. (2003). Bootstrap methods for time series. *International Statistical Review* 71, 435–460.

Harvey, A. C. and Koopman, S. J. (1993). Forecasting hourly electricity demand using time-varying splines. *Journal of the American Statistical Association* 88, 1228–36.

Hastie, T. and Tibshirani, R. (1990). *Generalized additive models*. London: Chapman and Hall.

Hastie, T. and Tibshirani, R. (1993). Varying-coefficient models. *Journal of the Royal Statistical Society – Series B* 55(4), 757–796.

He, X. and Shi, P. (1998). Monotone B-spline smoothing. *Journal of the American Statistical Association* 93, 643–650.

REFERENCES

Hor, C.-L., Watson, S. J., and Majithia, S. (2005). Analyzing the impact of weather variables on monthly electricity demand. *IEEE Transactions on Power Systems* **20**, 2078–2085.

Hyndman, R. J. and Ullah, M. S. (2007). Robust forecasting of mortality and fertility rates: a functional data approach. *Computational Statistics & Data Analysis* **51**, 4942–4956.

Kauermann, G. (2005). A note on smoothing parameter selection for penalized spline smoothing. *Journal of Statistical Planing and Inference* **127**, 53–69.

Kauermann, G., Krivobokova, T., and Fahrmeir, L. (2009). Some asymptotic results on generalized penalized spline smoothing. *Journal of the Royal Statistical Society – Series B* **71**(2), 487–503.

Kelly, C. and Rice, J. (1990). Monotone smoothing with application to dose-response curves and the assessment of synergism. *Biometrics* **46**(4), 1071–1085.

Kneip, A. and Gasser, T. (1988). Convergence and consistency results for self-modeling nonlinear regression. *The Annals of Statistics* **16**(1), 82–112.

Kneip, A. and Gasser, T. (1992). Statistical tools to analyze data representing a sample of curves. *The Annals of Statistics* **20**, 1266–1305.

Kneip, A. and Utikal, K. J. (2001). Inference for density families using functional principal component analysis. *Journal of the American Statistical Association* **96**(454), 519–532.

Kothandaraman, V. (1971). Analysis of water temperature variations in large rivers. *ASCE Journal of the Sanitary Engineering Division* **97**, 19–31.

Krivobokova, T. and Kauermann, G. (2007). A note on penalized spline smoothing with correlated errors. *American Statistical Association* **102**, 1328–1337.

Laird, N. M. (1978). Empirical bayes methods for two-way contigency tables. *Biometrika* **65**, 581–590.

Leurgans, S. E., Moyeed, R. A., and Silverman, B. W. (1993). Canonical correlation

REFERENCES

when the data are curves. *Journal of the Royal Statistical Society - Series B* **55**, 725–740.

Liu, J. M., Chen, R., Liu, L.-M., and Harris, J. L. (2006). A semi-parametric time series approach in modeling hourly electricity loads. *Journal of Forecasting* **25**, 537–559.

Liu, X. and Yang, M. C. K. (2009). Simultaneous curve registration and clustering for functional data. *Computational Statistics and Data Analysis* **53**, 1361–1376.

Loève, M. (1978). *Probability theory*, Volume 46 of *Graduate texts in mathematics*. New York: Springer.

Mammen, E. (1993). Bootstrap and wild bootstrap for high dimensional linear models. *Annals of Statistics* **21**, 255–285.

Marceau, P., Cluis, D., and Morin, G. (1986). Comparaison des performances relatives à un modèle déterministe et à un modèle stochastique de température de l'eau en rivière. *Canadian Journal of Civil Engineering* **13**, 352–364.

Mardia, K. V., Kent, J. T., and Bibby, J. M. (1979). *Multivariate analysis*. London: Acad. Press.

Martín-Rodríguez, G. and Cáceres-Hernández, J. J. (2005). Modelling the hourly spanish electricity demand. *Economic Modelling* **22**, 551–569.

McCullagh, P. and Nelder, J. A. (1999). *Generalized linear models*. Monographs on statistics and applied probability ; 37. Chapman & Hall/CRC.

McLachlan, G. J. and Krishnan, T. (1997). *The EM algorithm and extensions*. New York: Wiley.

Moral-Carcedo, J. and Vicens-Otero, J. (2005). Modelling the non-linear response on spanish electricity demand to temperature variations. *Energy Economics* **27**, 477–494.

Morrill, J. C., Bales, R. C., and Conklin, M. H. (2005). Estimating stream temperature from air temperature: Implications for future water quality. *Journal of Environmental Engineering* **131**, 139–146.

REFERENCES

Nielsen, H. A. and Madsen, H. (2000). Predicting the heat consumption in heating systems using meteorological forecasts. ENS. J. Nr. 1323/98-0025, Department of Mathematical Modelling, Technical University of Denmark.

Nielsen, H. A. and Madsen, H. (2006). Modelling the heat consumption in district heating systems using a grey-box approach. *Energy and Buildings* **38**, 63–71.

Ovidio, M., Baras, E., Goffaux, D., and Birtles, C. (1998). Environmental unpredictability rules the autumn migration of brown trout (salmo trutta l.) in the Belgian Ardennes. *Hydrobiologia* **371/372**, 263–274.

Ovidio, M., Baras, E., Goffaux, D., Giroux, F., and Philippart, J. (2002). Seasonal variations of activity pattern of brown trout (salmo trutta) in a small stream, as determined by radio-telemetry. *Hydrobiologia* **470**, 195–202.

Pagano, M. (1978). On periodic and multiple autoregressions. *The Annals of Statistics* **6**(6), 1310–1317.

Panagiotelis, A. and Smith, M. (2008a). Bayesian density forecasting of intraday electricity prices using multivariate skew t distributions. *International Journal of Forecasting* **24**, 710–727.

Panagiotelis, A. and Smith, M. (2008b). Bayesian identification, selection and estimation of semiparametric functions in high-dimensional additive models. *Journal of Econometrics* **143**, 291–316.

Paparoditis, E. and Politis, D. N. (2002). The local bootstrap for markov processes. *Journal or Statistical Planning and Inference* **335**, 959–962.

Pardo, A., Meneu, V., and Valor, E. (2002). Temperature and seasonality influences on spanish electricity load. *Energy Economics* **24**, 55–70.

Park, D., El-Sharkawi, M., Marks II, R., Atlas, L., and Damborg, M. (1991). Electric load forecasting using an artificial neural network. *IEEE Transactions on Power Systems* **6**, 397–402.

Pinheiro, J. C. and Bates, D. M. (2002). *Mixed effects models in S and S-PLUS*. New York: Springer.

REFERENCES

Rakowitz, G., Berger, B., Kubecka, J., and Keckeis, H. (2008). Functional role of environmental stimuli for the spawning migration in Danube nase Chondrostoma nasus (l.). *Ecology of Freshwater Fish* **17**, 502–514.

Ramanathan, R., Engle, R., Granger, C., Vahid-Araghi, F., and Brace, C. (1997). Short-run forecasts of electricity loads and peaks. *International Journal of Forecasting* **13**, 161–174.

Ramsay, J. and Silverman, B. (2002). *Applied Functional Data Analysis* (second ed.). Springer.

Ramsay, J. O., Bock, R. D., and Gasser, T. (1995). Comparison of height acceleration curves in the fels, zurich, and berkeley growth data. *Annals of Human Biology* **22**(5), 413–426.

Ramsay, J. O., Hooker, G., and Graves, S. (2009). *Functional Data Analysis with R and MATLAB*. New York: Springer.

Ramsay, J. O. and Li, X. (1998). Curve registration. *Journal of the Royal Statistical Society - Series B* **60**, 351–363.

Ramsay, J. O. and Silverman, B. W. (2005). *Functional Data Analysis*. New York: Springer.

Reinsel, G. C. (1993). *Elements of Multivariate Time Series Analysis*. New-York: Springer.

Rencher, A. C. (2002). *Methods of multivariate analysis*. New York: Wiley Interscience.

Rønn, B. B. (2001). Nonparametric maximum likelihood estimation for shifted curves. *Journal of the Royal Statistical Society - Series B* **63**, 243–259.

Ruppert, D., Wand, M., and Carroll, R. (2003). *Semiparametric Regression*. Cambridge University Press.

Ruppert, D., Wand, M. P., and Carroll, R. J. (2009). Semiparametric regression during 2003 – 2007. *Electronic Journal of Statistics* **3**, 1193–1256.

REFERENCES

Sailor, D. and Muñoz, R. (1997). Sensitivity of electricity and natural gas consumption to climate in the u.s.a.– methodology and results for eight states. *Energy* **22**, 987–998.

Sakoe, H. and Chiba, S. (1978). Dynamic programming algorithm optimization for spoken word recognition. *IEEE Transactions on Acoustics, Speech, and Signal Processing* **26**(1), 43–49.

Sargent, T. J. and Sims, C. A. (1977). Business cycle modeling without pretending to have too much a-priori economic theory. In C. A. Sims (Ed.), *New Methods in Business Cycle Research*. Minneapolis: Federal Reserve Bank of Minneapolis.

Schall, R. (1991). Estimation in generalized linear models with random effects. *Biometrika* **40**, 719–727.

Silverman, B. W. (1995). Incorporating parametric effects into principal components analysis. *Journal of the Royal Statistical Society - Series B* **57**, 673–689.

Smith, M. S. (2000). Modeling and short-term forecasting of new south wales electricity system load. *Journal of Business and Economic Statistics* **18**, 465–478.

Soares, L. J. and Medeiros, M. C. (2008). Modeling and forecasting short-term electricity load: A comparison of methods with an application to Brazilian data. *International Journal of Forecasting* **24**(4), 630–644.

Spearman, C. (1904). General inteligence, objectively determined and measured. *American Journal of Psychology* **15**, 201–293.

Steiger, J. H. (1979). Factor indeterminacy in the 1930's and the 1970's — some interesting parallels. *Psychometrika* **44**, 157–167.

Stock, J. H. and Watson, M. W. (2002a). Forecasting using principal components from a large number of predictors. *Journal of the American Statistical Association* **97**, 1167–1179.

Stock, J. H. and Watson, M. W. (2002b). Macroeconomic forecasting using diffusion indexes. *Journal of Business and Economic Statistics* **20**, 147–162.

REFERENCES

Stock, J. H. and Watson, M. W. (2006). Forecasting with many predictors. In G. Elliot, C. Granger, & A. Timmermann (Eds.), *Handbook of Economic Forecasting*. Amsterdam: North-Holland.

Tanner, M. A. (1992). *Tools for statistical inference*. Lecture notes in statistics ; 67. Springer.

Taylor, J. W. (2010). Triple seasonal methods for short-term load forecasting. *European Journal of Operational Research* **204**, 139–152.

Taylor, J. W. and McSharry, P. E. (2007). Short-term load forecasting methods: An evaluation based on european data. *IEEE Transactions on Power Systems 22*(4), 2213–2219.

Taylor, J. W., Menezes, L. M. d., and McSharry, P. E. (2006). A comparison of univariate methods for forecasting electricity demand up to a day ahead. *International Journal of Forecasting* **22**, 1–16.

Telesca, D. and Inoue, L. Y. T. (2008). Bayesian hierarchical curve registration. *Journal of the American Statistical Association 103*(481), 328–339.

The European Parliament / The European Council (2000). Directive 2000/60/EC establishing a framework for Community action in the field of water policy. *Journal of the European Communities*.

The European Parliament / The European Council (2005). Directive 2005/646/EC on the establishment of a register of sites to form the intercalibration network in accordance with Directive 2000/60/EC of the European Parliament and of the Counsil. *Journal of the European Union*.

Thurstone, L. L. (1947). *Multiple-Factor Analysis*. Chicago: University of Chicago Press.

Wager, C., Vaida, F., and Kauermann, G. (2007). Model selection for penalized spline smoothing using akaike information criteria. *Australian and New Zealand Journal of Statistics 49*(2), 173–190.

Wahba, G. (1990). *Spline Models for Observational Data*. Philadelphia: SIAM.

REFERENCES

Wand, M. (2003). Smoothing and mixed models. *Computational Statistics* **18**, 223–250.

Wand, M. P. and Ormerod, J. T. (2008). On semiparametric regression with O'Sullivan penalised splines. *Australian & New Zealand Journal of Statistics 51*(1), 179–198.

Wang, K. and Gasser, T. (1997). The alignment of curves by dynamic time warping. *The Annals of Statistics* **25**, 1251–1276.

Wang, K. and Gasser, T. (1998). Asymptotic and bootstrap confidence bounds for the structural average of curves. *The Annals of Statistics* **26**, 972–991.

Wang, K. and Gasser, T. (1999). Synchronizing sample curves nonparametrically. *The Annals of Statistics* **27**, 439–460.

Webb, B. W., Hannah, D. M., Moore, R. D., Brown, L. E., and Nobilis, F. (2008). Recent advances in stream and river temperature research. *Hydrological Processes* **22**, 902–918.

Weron, R. (2006). *Modeling and Forecasting Electricity Loads and Prices: A Statistical Approach*. Chichester: Wiley.

Wood, S. N. (2004). Stable and efficient multiple smoothing parameter estimation for generalized additive models. *Journal of the American Statistical Association* **99**, 673–686.

Wood, S. N. (2006). *Generalized additive models*. Chapman & Hall/CRC.

Wood, S. N. (2008). Fast stable direct fitting and smoothness selection for generalized additive models. *Journal of the Royal Statistical Society - Series B* **70**, 495–518.

Zitová, B. and Flusser, J. (2003). Image registration methods: a survey. *Image and Vision Computing* **21**, 977–1000.

Die VDM Verlagsservicegesellschaft sucht für wissenschaftliche Verlage abgeschlossene und herausragende

Dissertationen, Habilitationen, Diplomarbeiten, Master Theses, Magisterarbeiten usw.

für die kostenlose Publikation als Fachbuch.

Sie verfügen über eine Arbeit, die hohen inhaltlichen und formalen Ansprüchen genügt, und haben Interesse an einer honorarvergüteten Publikation?

Dann senden Sie bitte erste Informationen über sich und Ihre Arbeit per Email an *info@vdm-vsg.de*.

Sie erhalten kurzfristig unser Feedback!

VDM Verlagsservicegesellschaft mbH
Dudweiler Landstr. 99   Telefon +49 681 3720 174
D - 66123 Saarbrücken   Fax   +49 681 3720 1749
www.vdm-vsg.de

Die VDM Verlagsservicegesellschaft mbH vertritt

Printed by Books on Demand GmbH, Norderstedt / Germany